The
Dancing Girls
of Lahore

Fourth Estate

An Imprint of HarperCollins*Publishers*

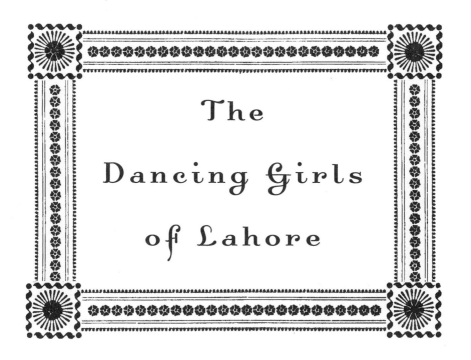

The Dancing Girls of Lahore

SELLING LOVE AND SAVING DREAMS
IN PAKISTAN'S ANCIENT
PLEASURE DISTRICT

Louise Brown

This is a true story. Although all of the characters are real, their
names have been changed to protect their privacy. The only real
names used in the book are Sheikh Zayed, the former president of
the United Arab Emirates and ruler of Abu Dhabi, now deceased;
Iqbal Hussain, the artist who paints portraits of the women of Heera
Mandi; and Tariq Anayaat, the sweeper.

HarperCollins books may be purchased for educational, business, or
sales promotional use. For information, please write:
Special Markets Department, HarperCollins Publishers,
10 East 53rd Street, New York, NY 10022.

FIRST EDITION

Designed by Claire Naylon Vaccaro

Printed on acid-free paper

Library of Congress Cataloging-in-Publication Data

Brown, T. Louise, 1963–
The dancing girls of Lahore: selling love and saving dreams in
Pakistan's ancient pleasure district/Louise Brown.
p.   cm.
ISBN 0-06-074042-6
1. Women—Pakistan—Lahore—Social conditions.
2. Prostitution—Pakistan—Lahore.   3. Prostitutes—Pakistan—
Lahore.   4. Lahore (Pakistan)—Social life and customs.   I. Title.

HQ1745.5.Z9L333   2005
306.74'2'095491—dc22   2004063596

06 07 08 09 BVG/RRD 10 9 8 7 6 5

For Maha

Contents

Contents

Prologue

Five new girls are staying in the thin pimp's brothel. The bold ones come to the door, laughing and pulling their veils over their hair as they glance around the courtyard. The others stay in the shadows, edging closer only occasionally to peep around the shutters or lift the bamboo blinds. Like dozens of girls who have passed through this brothel, they will spend most of their day in the damp, windowless interior of the house.

The shrine looks as if it has always been in the corner of the courtyard: as if the devout have lit oil lamps and prayed beneath the Shia banners for centuries; as if the straggly tree and the bushes have always grown there and have always been strung with pretty lights on every religious occasion. Traditions must be swift to take root in Heera Mandi because, five years ago, there was no shrine; it was the place where the pimps relaxed on wood-framed, rope-strung beds when it was too hot and humid to sit inside their den.

Another family has moved into Maha's rooms and a group of Afghans have set up a miniature refugee camp on the roof using bits of rope and ripped blankets that are permanently sodden with the winter rain. Her plants have gone from the balcony and a new collection of washing is drying slowly on the railings. The giant air cooler no longer juts precariously out of the window, threatening to crush the passers-by below. There's no singing from the second floor of the big yellow house. Maha's voice has stopped echoing round the courtyard as she practices her *ghazals*, and the musicians have ceased carrying their *tablas* and harmonium up the narrow spiral staircases to her rooms.

When I first came to the courtyard, things were very different. The cycle of life has spun quickly, occasionally with cruelty, usually with bitter inevitability, and sometimes with such fast-burning beauty and energy that a single moment of brilliance illuminates whole lives in the dark, hidden world of this ancient brothel quarter.

A rickshaw draws alongside me and a hand decorated with gold filigree rings beckons through the fractionally opened door. Inside, one of the passengers lifts her veil. My friend's eyes are smiling at me from a girl's face: Maha's daughter has blossomed into a stunning young woman.

"Louise Auntie, *chale*, let's go," she requests with all the sweet charm she had as a child. "We've been waiting for you."

"We Were Artists . . . Not *Gandi Kanjri*"

(Hot Season: April–June 2000)

Lahore is a wonderful city with rich character and a worn charm. The Mughal Empire
has bequeathed some glories to the modern
city: the awe-inspiring Badshahi Masjid;
the imposing Shahi Quila, or Royal Fort;
the pretty Shalamar Gardens; and the now
dilapidated tombs of Emperor Jahangir and
his empress, Nur Jahan. Grand buildings
inherited from the British raj sit in stately,
shabby order on the broad, leafy Mall
Road running through the center of town.
New suburbs have grown—some affluent
and some not. The streets and markets bustle and hum with life and the mosques and
mausoleums are always busy. Best of all,
though, is this ancient place—the Walled
City—a quarter of a million people squeezed
into a square mile of congested tenements
and shops. It is the heart of Lahore and it
carries the city's soul.

Old Lahore can't have changed much for centuries. The moat was filled in long ago and the defensive walls have gone, but the residents, constrained by ancient land boundaries and historical memory, continue to build their houses as if the walls still exist: an ageless and invisible presence. The thirteen gates into the city remain too, channeling pedestrians and traffic from the wide roads of contemporary Lahore into the narrow lanes and alleys of the Walled City. Rickshaws, horse-drawn carts called *tangas*, motorbikes, and small vans compete with pedestrians for space inside the walls. No vehicles of any kind enter the narrowest alleys. Neither does the sun. Only in the wider lanes and the bazaars does the sun shine directly on the ground. Most of the small passages running through the city lie in perpetual, dusty gloom.

Early morning is the best time to see the old city. During the hot season there are a couple of hours before the temperature soars and the lanes become too congested. The city wakes up and life unfolds in much the same way it must have done hundreds of years ago. The shopkeepers are busy: the butchers slice up chickens and goats, the tea shops open and the bakers prepare *halva* and fry *puri* for the first meal of the day. The fruit and vegetable sellers arrange their produce in a kaleidoscope of bright colors: plump aubergines, *mooli*, red carrots, sweet firm tomatoes, bundles of spinach, fresh okra, and leafy bunches of coriander and mint. Donkey carts rattle and creak down the *galis*, the narrow lanes, delivering goods: large round metal pots carrying milk from the villages; another piled high with sacks of flour and rice. A rickshaw whose only passengers are a dozen frantic hens stops and the goods are thrown, squawking, into the back room of a butcher's shop. In the little workshops men and boys are already at work by seven o'clock, grinding bits of metal, heating syrupy liquids over open fires, sticking unidentified items together. It is gray, dirty, repetitive work and it lasts for most of their waking day.

Heera Mandi—the Diamond Market—is a crumbling ghetto of three- and four-storey buildings tucked into the northern corner of the Walled City, right next to one of the greatest forts of Mughal

India and its biggest, most perfectly proportioned mosque. The old women living here say it has been the red-light district for as long as they can remember and it flourished long before the British arrived in the mid–nineteenth century. Heera Mandi, also known as Shahi Mohalla, was important then, and in its heyday it trained courtesans who won the hearts of emperors. The old ladies insist that things used to be different in those times: women like them were respected. They were artists, not *gandi kanjri*—not dirty prostitutes.

I have a room in the home of Shahi Mohalla's most famous resident, Iqbal Hussain, a professor of fine art who paints portraits of the women of Heera Mandi. When I came to Lahore previously it was Iqbal who taught me most about prostitution in Pakistan and about life in the *mohalla*. He is an authority on the subject because he lives and breathes it: it's in his blood. He is the son of a courtesan and has spent over half a century in Heera Mandi, growing up in this house that lies in the shadow of the mosque and in the longer shadow of social stigma. His friendship gives me some protection now that I've returned to stay in the *mohalla* and witness its life first-hand.

Iqbal's house expands, month by month, as he scours the construction sites of the Walled City, collecting windows, doors, statues, and tiles from ancient, demolished *havelis*—the graceful traditional homes of the rich. He incorporates these fragments into his home, so it has become an eclectic fusion of Hindu, Muslim, and Sikh design. My room, on the third floor of the house, overlooks the biggest courtyard in Heera Mandi. It's the most beautiful room. It has three bay windows, each fitted with tiny panes of colored glass. The furniture and doors are of carved wood and the giant floor cushions, bolsters, and heavy curtains are made of golden and burgundy brocades. This room, like the whole house, has been assembled from pieces and images of old Lahore.

On the ground floor of the house Iqbal runs a restaurant where young couples meet for forbidden romantic liaisons during the afternoon. They sit in the back room and drink bottles of 7-Up in the summer and cups of coffee in the winter. The boys talk a lot and the girls giggle without reason or pause. In the evening most of the visitors are groups of well-heeled, arrogant men. At other times entire families come for an outing bringing Grandma, the babies, and assorted uncles and aunts. They dine at long tables and then traipse up to the roof to look at the Badshahi Masjid and the fort. As they pass my room I hear them puffing and complaining that the climb is steep and that there are a crippling number of steps.

There's something exciting and illicit about coming here, something that makes respectable Pakistani pulses race. They park their air-conditioned cars right outside the restaurant, rush inside, and, after their meal, peep into the courtyard—into the dangerous scandal that is Heera Mandi.

From one side of Iqbal's roof terrace you can see right into the heart of the marble-domed Badshahi mosque. All day and well into the night, straggly lines of barefoot men make their way to prayer across the vast quadrangle. At dusk, a couple of hundred boys and youths play cricket on the field by the mosque and men sit in circles deeply involved in a debate. A group of heroin addicts crouch at the edge of the grass where a copse of trees gives some shelter and privacy.

Lining the opposite horizon are the rooftops of the old city's unplanned and ramshackle houses. Black Shia Muslim flags with red fringes are strung on rusty poles, and metal *panja*—the mystical hand that is the symbol of Shia Islam—rise high above the buildings. The roofs are a mess of steps, crumbling bricks, powdery cement, and terraces cluttered with all kinds of debris: piles of rugs, blankets, and bits of old furniture. The day's washing is draped over walls and shutters, and here and there, almost lost in the chaos, are carved wooden doors, trellises, and ornate plasterwork—reminders of a more prosperous Heera Mandi when Lahore was still a multifaith and multicultural city.

Forty years ago Heera Mandi was ornate: the old buildings lining the main roads had exquisitely worked *jharoka*—finely carved, wooden bay windows—and balconies. Today, very few remain. They have been torn down and replaced by ugly concrete blocks with simple wooden shutters or crude metal grills. The revamped buildings may be more practical to live in, but they have none of the allure of old Lahore.

Men rest on the roofs in the late sun. A few women sit with them, combing through their children's hair in search of lice. Other children lean precariously out of windows and over walls, looking into the courtyard where a couple of fat, elderly women recline on *charpoys*, sagging, wood-framed rope beds, gossiping and chewing and spitting out their *paan*. A dozen little boys race around them playing games and fighting over a tricycle. These children live in the houses surrounding the courtyard. Half a century ago some of these buildings must have been grand residences, but today they have been divided and subdivided into one-, two-, or three-room apartments. Other houses that line the narrow alleys must always have been oppressive tenements with little light and no fresh air.

By dusk the rats run and jump in a fast-moving gray stream from one building to the next. The last of the light slips behind the domes of the *masjid* and the *azan*, the call to prayer, begins. A woman sings a *ghazal* in one of the houses opposite, her lovely voice reverberating around the courtyard until she too hears the *azan* and grows quiet. She sings every day in the shadows of her room, and we catch a glimpse of her beauty when she passes the window. Sometimes she is with her children and sometimes a man comes to visit. It's always the same man, and when he beats her we pretend not to see.

I love this part of the day on Iqbal's roof terrace, four storeys above the street. I sit in the twilight, jotting ideas and images into a big, untidy notebook. There is much to write about, so much to see and understand that I fear I'll only be able to capture glimpses of this forbidden subculture. I've spent the last five years of my professional life as an academic researching prostitution and the trafficking of women in Asia. I've looked at human rights issues, debt bondage, and HIV/AIDS in locations as diverse as the clubs of Tokyo, the pedophile haunts of Phnom Penh, the girly bars of Bangkok and Manila, and the giant brothels of Mumbai, Calcutta, and Bangladesh, so large and heavily populated that they form whole subsectors of the city. Heera Mandi is like these places, and yet it is not. It still retains elements of India's traditional pleasure quarters, but it is changing fast and I have come here to record these changes; a witness to the closing of an era. I'll stay in the *mohalla* for a month or so, two or three times a year. Visiting for longer will be difficult because I have a job to do at home, teaching in a university, and I have three children: a boy and two pretty girls just the right age for the business. I can't bring them to Heera Mandi to stay in the dubious safety of seclusion, so they'll remain in England with my mother.

On the far side of the courtyard a couple of young women are folding dried bedsheets, laughing and moving playfully in opposite directions so that the sheets twist into a rough plait. I stop writing to watch them. Their younger brothers ignore them, too preoccupied with their kites and the rough skirmish to reach the highest point of the building. When the afternoon and evening weather is fine, dozens — sometimes hundreds — of kites flutter above the old city, tiny specks of color swooping and soaring in a dainty, pretty display. It looks such a gentle, well-mannered hobby, but in reality each kite flier is locked in deadly combat to send their rivals' kites plummeting to earth.

On the roof of a neighboring house children somersault over a bulky roll of ancient bedding. An old lady — their grandmother perhaps — sits in a corner observing the game while patiently picking sticks and tiny stones out of the lentils she's spread on a giant tray. The vision of Heera Mandi you see from the streets is only a partial view of the *mohalla*. There's another world inside the buildings, the hidden world of the women, and there's activity above the buildings too: the slower but ceaseless life of the inner city's rooftops.

The courtyard wakes up at night and its lights remain on until dawn. Motorbikes roar in and out; the popcorn seller wheels in his musical cart; the pimps congregate on a couple of *charpoys*; a dog barks and triggers a dozen others. For women in this part of the city the world is largely nocturnal. They're at home but they're not asleep. They lie on mats or curled up on beds. They watch television, or listen to music while they eat and drink. They're waiting. Every now and again a house closes its shutters. An hour or so later they will open again. Men walk around in small groups of two or three. They disappear into pitch-black alleys that are hard to negotiate even in the daytime. They must have been here before to be so sure of their step.

The front of Iqbal's house is on Fort Road. The road runs along the perimeter of the mosque and the fort, and a hundred meters from his door is Roshnai Gate—the Gate of Splendor or the Gate of Light—so called because it used to be brilliantly illuminated. This is where the people of the old city—what the local people call the *andron shaher,* or inner city—enter the Hazoori Gardens. From the gardens they can turn into the entrance to the Badshahi Masjid or through the Alamgiri Gate into the fort. Once, this was the focus of power, not just of the city but of an entire empire too.

Lahore is the cultural center of the Punjab, one of South Asia's most important and wealthy regions. The Punjab derives its name from the five rivers, or *panch ab,* which draw together to create an agricultural plain and a power base from which rulers have controlled large parts of the subcontinent. During the Mughal era, from the sixteenth to the eighteenth century, Delhi, Agra, and Lahore were all, at different times, the imperial capital and the emperors constructed fabulous buildings in the cities to celebrate their religion, their power, and their women: the Taj Mahal in Agra; the Red Fort and the Jama Masjid in Delhi; and the Royal Fort and the Badshahi Masjid in Lahore. Mughal emperors paraded on their elephants through the Alamgiri Gate of Lahore's fort, and their courtesans, their dancing girls, and their courtiers came this way too, along with merchants, jewelers, tailors, and scribes, after passing first through the sparkling Roshnai Gate: the great doorway linking the people with their ruler and their place of worship.

Roshnai Gate is not so glorious today, although I doubt whether life for those dancing girls slipping through the gate could ever have been quite as romantic as it is in legend. The most attractive might have won a place in a harem as one of an aristocrat's many concubines, but for the others, life must have been much the same as it is now: a short life of flattery and a much longer life in retirement.

Elephants no longer wait outside the Alamgiri Gate: only cars and vans transporting tourists. But our side of Roshnai Gate—the inside—must have looked little different a century ago. It's a lot more polluted now, and there are more motorized rickshaws than carriages, but the snack *wallas* are still there with their carts, still alternating their treats according to the season: syrupy sugar cane and mangoes cooled in ice-filled buckets in summer and hot, charred, sweet potatoes to warm us in the winter chill. The corner stalls sell Coke and Sprite today rather than delicious *sherbet,* the old-fashioned sweetened, diluted fruit juices, but the Al Faisal Hotel, just next door, serves up good traditional fare cooked in big pots over a fire. The favorite dishes are *dal,* vegetables, lamb or goat in a tasty curry gravy, or *salan,* and wonderful thin unleavened breads called *rotiya.* The baker mixes flour and water in metal buckets, rolling and slapping soft balls of dough into rounds that are pressed against the sides of a kiln-shaped oven only to be flipped out a few moments later, puffed and scorched at the edges, and so hot you can't hold them. Most of the customers eat on the premises, sitting at trestle tables inside the open-fronted restaurant. Almost all the diners are men, but there's a little corner for women too with a flowered curtain that can be pulled around so they can eat in the comfort of *purdah,* seclusion. It looks like a shower cubicle.

It's very rare for the flowery curtain to be used because there are hardly any young women about. The few who are in the streets are veiled and walking fast. Girls and women between the ages of 12 and 50 stay inside their homes for their protection and to preserve their honor. Most of their lives are spent in a handful of rooms. They don't even go out to do the daily shopping: it's the boys and men who run errands.

Heera Mandi is an excellent place to buy food and the tea shops and restaurants are invariably packed with men buying cups of sweet milky tea, lamb kebabs, irresistible barbequed chicken, freshly baked breads, and crisply fried snacks that are mouth-wateringly tasty. Such delectable snacks also have an insidious and profound effect on the contours of the body. The poor of all ages are thin and pinched, but the more affluent young men who live in Heera Mandi, or who visit it for fun, are strikingly handsome: dark, with strong features and a confident manner. By the time they are 30, they are stout. They've had too many of those fried snacks. By 40, many have round, puffy faces and waddle like heavily pregnant women.

Fort Road turns by Roshnai Gate and runs along the wall of the fort. Workshops and garages cluster here: small colorless places where they fix rickshaws, motorbikes, and cycles. A dozen barbers squat on the pavement giving haircuts and shaves with lethally sharp razors. A few specialize in ear treatments and use miniature spoons on long thin handles to furrow around and extract clods of orange earwax from clients who seem to enjoy the dig.

The character of the road changes farther along: the shops are bigger and more prosperous. In the afternoons, one of the bakeries sells the most perfect *samosas* and a delicious, aromatic bean soup thick with finely chopped coriander. Customers form an untidy queue and then dash home with their just-fried *samosas*, the oil from the pastry soaking through the paper wrapping and the hot soup slopping in knotted plastic bags. Heera Mandi Chowk —*chowk* meaning intersection —lies at the top of the road, and from here, you can move quickly to more respectable areas through a road lined with shops that make and sell traditional musical instruments and dental surgeries that advertise their services with large, macabre, hand-painted illustrations of dissected mouths and heads.

These days, Heera Mandi's main street is full of tea shops, restaurants, and small rooms with pool tables or pinball machines with multicolored flashing lights. By Tarranum Chowk a cinema promotes its films with pictures of plump, pretty girls on a big colorful billboard. It's the busiest, most congested part of the *mohalla;* in front of the cinema, speeding rickshaws, carts, motorbikes, and pedestrians are all pulled into a frightening vortex of spinning bodies, donkeys, and metal, only to be spat out again, usually unscathed, to continue their journey through Heera Mandi.

These streets are never quiet and the *galis* and alleyways that criss-cross them are rarely empty, even at night. *Especially* at night. During the day Shahi Mohalla's rundown streets look like any other part of the Walled City, but at dusk, this part of old Lahore transforms itself. The restaurants are packed. The revelers—all men—saunter hand in hand. It is noisy, lively, and exciting. Men jostle in the alleyways and the bazaar is jammed with traffic. Popcorn and ice cream *wallas* push musical carts through the crowds, maneuvering around potholes and mounds of rank garbage. The walls of an office at the corner of the main street are plastered with photographs of actresses. It's not clear what services the office is providing, but it's doing a brisk trade. There is a sign in the window in Urdu, and an attempt at an English translation, which reads, "Best Music Gorup—Eny Fungton."

By nightfall a handful of middle-aged women sit by open doorways discreetly looking for clients. Up above street level and above the still-closed performance rooms, younger women move on dimly lit balconies. This is where the real business of Heera Mandi is done. A single young woman walks in the street—a tall, striking figure swinging through the crowd with an exaggerated sexy walk. She's not veiled and she's wearing a red *shalwaar* and a red-and-white flowered *kameez*. Tossing her hair and holding her head high, she is bold and confident. I look at her again and understand why: she's not a woman, she's a *khusra*—half man, half woman.

In the day you would never guess that Heera Mandi is a red-light district in which hundreds of women live and work. Prostitution is illegal in Pakistan and so is all sex outside marriage. *Zina*—extramarital sex—is punishable under Islamic law by stoning to death or by a prison sentence and a brutal whipping. Women who sell sex are the lowest, most vilified women on earth. But every day, for an hour on either side of midnight, Heera Mandi makes a concession to lewdness. A couple of dozen shops open their shutters and the women inside aren't veiled. They wear elaborate, brightly colored dresses, great quantities of makeup, and sit on sofas and chairs that are draped in shiny satin covers. Their smiles are strangely stiff, but that's because they can't see into the darkness of the streets. Expensive cars pull up. Sometimes the men come in to buy a dancing show. The shutters are closed with a loud clatter, the sofas are pushed against a wall, cold drinks are brought for the men, white sheets are spread on the floor, and the musicians sitting in the corner of the room begin to play their harmoniums, *tablas*, and *dholaks*. These are the *kothas*, or performance rooms, of Heera Mandi, and they are where the best courtesans of the *mohalla* have traditionally made their name.

Maha

Maha lives with her five children in an apartment on the second floor of a house lining the big, open courtyard. I've caught fleeting glimpses of her many times from my windows. Today, she's called me in to visit and is curled up on a surprisingly expensive-looking sofa. The scene could be respectably middle-class but for the stench from the drains and the rubbish strewn on the grim spiral staircase leading to her rooms. She's lovely, with a natural poise, and her long thick hair is hennaed to a dark, glossy red. She's plump but still graceful, and her dancing is superb. Ten years ago, before the pounds piled on, she must have been astonishingly beautiful. Now she's in her midthirties and the mother of too many babies. Her children are delightful but they've been a disaster for Maha's career: they've not done her figure or her finances any favors.

Her mother and sister live downstairs. Neither woman is blessed with Maha's beauty. The older woman sits at the window for most of the day, grinning and chewing *paan*; she's not always friendly. Maha says they nag her because she's no longer bringing in money. Her mother insists that it's such a waste for Maha to stay in her rooms when she retains enough beauty to enchant the customers in the bazaar.

Maha refuses to dance in the *kothas* of Heera Mandi because she says that Adnan, her husband, will disapprove. Maha is Adnan's second wife and more like a mistress than a wife. Muslim men can have up to four wives at any one time and these can easily be replaced through *talaq*—a quick divorce. Adnan has withdrawn most of his financial support from Maha and she thinks the divorce will follow shortly.

Maha is lucky to have a legal marriage: most women in Heera Mandi never experience such a thing. They still, though, call their clients *shohar*, husband, because otherwise it would mean confessing to a criminal act. Indeed, a very large proportion of the women incarcerated in Pakistani jails are there because they've been convicted of having sex with someone who is not their husband. Not surprisingly, there is not an equivalent number of men imprisoned for having sex with a woman who is not their wife.

Adnan disapproves of many things. Five years ago things were different. He was a successful businessman who installed Maha in a nice house outside the walls of the old city. She has two children by him: Mutazar, who is 4 and the only boy in the family, and Sofiya who is 18 months old. Now, though, her happy life is changing. Adnan is growing tired of Maha and tired of supporting her and the children. Maha weeps and her nose turns pink. "Adnan loved me but after I had so many babies he told me to get out and go back to Heera Mandi. He doesn't love me anymore because I'm fat. I'm old and finished." She's probably right. Maha's story is a common one: pretty women from Heera Mandi win a temporary reprieve from the brothel in their twenties only to return in their thirties. Maha has come back to the place where she was born and has always belonged.

We watch a video shot at the wedding of one of Adnan's nieces, that is, a niece from his official, respectable family. Maha wasn't invited to the wedding, but she has made up for her absence at the auspicious occasion by viewing the video so often that it's almost worn out. She knows every frame. "There's Adnan," she shouts, pointing at the television. Adnan is twenty years older than her and seems a friendly kind of chap, stooping slightly and grinning hesitantly at the camera. Mumtaz, his wife, has a blank-eyed smile and a stunning collection of jewels.

Maha pauses the video so we can analyze the guests in greater detail. "Mumtaz doesn't have a nice nose," she states. "If a woman has a bad nose she's no good. And her eyes are not as nice as mine, are they? She's forty-three. She thinks she's a sexy lady." She begins to laugh but it tails off into more crying. I agree: Mumtaz is neither as beautiful nor as young as Maha.

maha's new, much inferior house has two rooms. One of these small rooms is dominated by an ancient air cooler that expels a blast of air so ferocious it almost rips the hair out of your scalp. It creates even more noise than it does breeze, and whenever guests arrive during the hot season we are forced to shout over the clattering and whirring. The other piece of furniture is a mattress on which I spend hours watching the life of the courtyard as Sofiya, Maha's tiny daughter, who is all fat little thighs and snot-encrusted nose, tramples over me, sharing my drinks, chewing the straws until they are useless, and then finding bits of old food on the mattress to put into my mouth.

The courtyard is a lively market and a constant flow of salespeople pass through. *Sherbet* sellers crush ice and mix it with artificial fruit syrups to drink on hot days, and a man with a pushcart sells every type of cheap clip, elastic band, and plastic accessory that has ever been designed for ladies' hair. Fairground rides on squeaky metal wheels rattle into the courtyard: swings and miniature Ferris wheels, all trailed by a long line of excited children. A good trade is done in secondhand clothes imported from abroad and heaped on carts. Some of the other traders carry their wares around with them: two dozen plastic bomber jackets; an armful of enormous cream-colored cotton gloves that will fit no one; and a bamboo pole strung with garlands of flowers. The garlands are fragile and beautiful, but they last only a few hours before wilting and disintegrating into a handful of delicate, brown-edged petals. Some are made from *gulab:* pretty pink flowers that look like wild thornless roses. Others are made from *motiya:* white flowers, like a tiny bud, with an unforgettable, heady fragrance. Maha keeps the garlands in little dishes in her room to scent the air.

Some of the visitors to the courtyard are selling services. The *malish karne wallas* sell massages, advertising their skills by rattling bottles of oils on a metal tray. The clinking bottles sound like castanets. A barefoot man with a big gray beard sings religious songs praising Allah and asking the good householders for a few rupees. He has a strong, distinctive voice and he tours the courtyard every day or two singing the same tunes. The occasional rupee note flutters down to him from a window and he moves on, his voice becoming fainter and fainter as he sings his way down the *gali.*

Maha and her daughters are attentive hostesses, forever presenting me with a succession of snacks: *namkeen,* a spicy fried snack of wheat, nuts, and *dal;* potato chips; green mango with chili; biscuits; 7-Up and Coca-Cola. In Muslim societies it is customary to show kind and generous hospitality to guests, and visitors to Pakistan will never go hungry. The welcome is always warm, and the poorest household will give their best, even if this means that they themselves will go without. It's a generosity I often find embarrassing, sometimes because I know my hosts cannot afford such kindness and sometimes because I visit lots of houses each day and am obliged to eat in every one. The food tastes wonderful but everything is fried. The meat swims in *ghee* and the fizzy drinks taste sweeter than at home. I put on seven pounds each month I stay here. No wonder the women of Heera Mandi look like they do: young women here are lithe and pretty, but fifteen years later most are obese. Lives bound to a few rooms, an unhealthy diet, and a complete absence of exercise results in atrophied muscles and generous layers of fat.

Maha and her relatives are interested in my size and surprised that I'm 36 years old and have three children. "Where is your belly?" they ask. "What happened to your hips? Your hips are very small." Maha and her cousin look at me with compassion because it's a well-known fact in Shahi Mohalla that men prefer women with ample hips.

Maha's two older daughters, Nisha and Nena, are playing in the room. They are children from Maha's first long-term relationship. When she was 15 she was taken as a mistress by a *sayeed*. A *sayeed* is a Muslim who can trace his lineage to the family of the prophet Mohammed. Maha thinks her relationship with a *sayeed* endowed her with honor and she repeats his full name over and over again so I can absorb its significance. Some of his social capital has rubbed off on her, and she's immensely proud that her children have an important father. It endows her with status by association: these children haven't been fathered by an unknown client.

Long before the *sayeed*, Maha was connected with other distinguished men, the kind of men that only the most beautiful and accomplished dancing girls can claim as patrons.

"My very first husband was very important and very old," she gushes. "He was Sheikh Zayed. He married my sister, Fouzia, a few weeks before he married me, but he was angry because she didn't bleed."

Poor Fouzia was sent home in disgrace. The sheikh liked virgins, and young ones at that. As compensation, Maha was dispatched to Karachi to meet the great sheikh.

"He paid so much for me, two *lakh*"—200,000 rupees ($3,372)— "and I was only twelve." Maha regards it as an honor, and maybe it was: Sheikh Zayed was the ruler of Abu Dhabi and the president of the United Arab Emirates.

She can't remember much about the event because she was sedated, but she knows she wasn't with him for long: perhaps only an hour. Then she was taken to another man—someone much younger and much more handsome. He liked her a lot and kept her in his bed for a month. She laughs whenever she repeats the young sheikh's words. "He told me, 'You are a very sexy girl.'"

Nisha and Nena don't share Maha's sensuality or her magnetic sexuality: they are eclipsed by her—pale and pretty, half-grown shadows of their mother. Nisha, the older girl, is tall, thin, and quietly angry—laughing and yet simmering with a resentment I can't quite place. Her younger sister is softer, more compliant, her wide, long-lashed eyes frequently cast down with a childish shyness. She is very kind, very dutiful. One day, when I felt ill, Nena insisted on fanning me as I fell asleep on the mattress. When I woke, two hours later, she was still there, still smiling and wafting the fan.

I can't imagine that these girls will make successful prostitutes. Their fate, though, has been sealed from birth. They are barely literate. They don't go to school. In fact, they don't go anywhere. They spend their lives in these two dark rooms in the corner of the courtyard, tripping down the spiral staircase, hovering around the entrance to the alleyway, and occasionally going in a rickshaw to the bazaar to buy food and clothes. That is the extent of their world.

In the life cycle of traditional Heera Mandi, Maha's family would be poised to transform itself. The *tawaifs*, or courtesans, of old Lahore withdrew from selling sex in their early thirties because there was no longer any real and profitable demand for them. Only desperate common prostitutes continued selling services into middle age. Refined courtesans with honor to preserve went into graceful retirement and managed the careers of their daughters and nieces. A girl who gave birth to a daughter when she was 15 would have someone to replace her in the business when she was 30. Giving birth to a girl was like producing your own personal pension plan because a daughter's youth and beauty sustained her family. This transition has not yet happened in Maha's household, but for Nena and Nisha—now 12 and 14—there will be no escape from the bazaar. In Pakistan, marriage is still most women's only option in life, and, unluckily for Nisha and Nena, no one will seriously consider marrying the daughter of a prostitute. As Maha so often reminds me, the daughter of a dancing girl always becomes a dancing girl. They pass the occupation and the stigma from one generation to the next like a segment of DNA.

Girls in Heera Mandi grow up in a completely different environment from ordinary Pakistani girls. In the *mohalla*, female beauty and sexuality are openly celebrated. From the time they are babies, girls witness a stream of men coming to the doors of their mothers and aunts and know that, when they grow a little older, these men will visit them too. Not all live easily with this, but most do: there is no alternative, and within the enclosed world of the *mohalla*, it is not considered wrong or bad. Indeed, those who perform the task well, expensively and with dignity, are lauded and envied.

Maha spreads herself out on the mattress in a lilac *shalwaar kameez*. Even when she is sitting down, she is loud and energetic. Adnan is about to visit and she's preparing herself with concentrated enthusiasm. She has sprayed perfumed deodorant everywhere, drawn thick black lines around her eyes, brushed her hair, and colored her lips in vivid fuchsia. In between the preparations she shouts at the children, cuffs her son on the head, and complains that Adnan's scheming wife—his official wife—is sabotaging her place in his heart.

She orders the children out of the room, takes my hand, and whispers, "Mumtaz, that bitch woman, said to Adnan, 'Why do you like Maha when she has had so many children? She is big . . . big.'" Maha gestures to her groin, makes a wide stretching motion with her hands and performs some realistic grunts to mimic giving birth. She's serious: in Heera Mandi having children is bad for business, at least in the short term. Babies prove your secondhand status and suggest that you have a slacker, and cheaper, pelvic floor. The most successful of the elite courtesans may never have children because of the impact on their figures, and in old age they will live off their savings and the earnings of their nieces.

Maha hisses at me that Mumtaz is keeping her husband at home by offering him an unlimited diet of oral and anal sex. Maha makes clear that doing it in the *bund* (anal intercourse) is something she neither likes nor is capable of performing.

I ask her how she can be so sure that Mumtaz is offering these sexual favors.

"She's a Pathan," Maha shouts. "They all do it."

In Heera Mandi, Pathans—also known as Pashtuns or Pakhtuns—are treated like a joke, and they are reputed to have a fondness for anal sex. Pathans come from a tribal society spanning the borders of north Pakistan and Afghanistan. Those in Swat in north Pakistan practice strict seclusion of their women. A girl is said to enter her husband's house in a bridal dress and to leave it in a coffin. No girls and young women are visible in the villages and towns of the area because they are all inside their homes. Outside, in the men's world, young men and women have no contact, and in the absence of female sexual partners, frustrated men turn to teenage boys and sometimes to those who are even younger. Beardless boys with soft skin are highly prized. Until very recently it was a sign of status for a man to keep several *bedagh* (passive male partners) to cater to his sexual needs: it was considered entirely normal. Even today, most young men's first sexual experiences will be with a *bedagh* or with a friend, and sex with boys continues to be considered less demanding and more pleasurable than sex with women.

The next day Maha's eye is swollen and bruised. Adnan paid her a visit but the eyeliner and lilac *shalwaar kameez* were wasted on him. She hangs over a bowl bathing her face with iced water and cries, "I want to die. I want to cover myself in kerosene and light a match. Then I'll die and be happy." She scrambles around the room hunting for matches only to find that there's barely a drop of kerosene left in the plastic bottle. She has no intention of carrying out the threat, but she wants some sympathy. For the whole afternoon she sits by the window holding a tissue to her eye and waiting for Adnan to drive into the courtyard.

When he doesn't arrive we decide to search for him. Maha spends a long time disguising the bruises before we run to find a rickshaw, our veils flapping and our sandals slowing our progress. Adnan owns extensive property in Lahore. It's not prime real estate, but there's a lot of it: streets and streets of lower-middle-class housing and small businesses. Maha points out the house she lived in before she was packed off back to Heera Mandi. It's a fairly ordinary-looking place but it's a palace in comparison with her home in the courtyard. We spot Adnan ambling along the street, and Maha tells the rickshaw *walla* to stop. Maha keeps glancing at herself in a mirror fixed inside the rickshaw, adjusting her hair and the drape of her *dupatta* on her head and shoulders. She pleads with Adnan and then argues with him because he's wearing a heavy gold chain that his wife has bought for him. Adnan smiles and laughs as if it's a big joke. "Tell him I love him," she says to me in strangled sobs. "Tell him his wife is old and I am more beautiful. Tell him sex with me is better than with Mumtaz."

Children of the Mohalla

I haven't seen Maha and the children for three weeks. They've moved into a new house. It's a modern place, but the bedroom has no windows, and although there's a bathroom with indoor plumbing, the room has no ventilation and there are frequent floods because the plumbing disconnects itself and water from the sink and the bath gushes out onto the floor. Going to the toilet is like going paddling. The new residence does have one advantage: Maha runs her fingers lovingly over a couple of plastic cupboards and sighs, "It's an American kitchen." This means you have somewhere to put your pans and bits of food rather than simply piling them on the floor.

There's been a marked change in Nisha, the oldest girl. She's much thinner and is lying curled up on the bed shivering and looking hot. She rolls off and staggers into the other room giving me a wan smile.

Maha is in such deep and angry despair that she doesn't notice.

"Some days Adnan doesn't visit and I've no money," she complains. "He gives me three hundred rupees [$5] on the days he does come and sometimes there's not enough money to feed the children. I can't go on like this." She stomps around the room shouting. "I keep praying for that bitch Mumtaz to die, but nothing happens. So I'm going to Bahrain with a group. I'll earn lots of money. I'll sing and men will look at me and give me jewelry. I can earn two or three hundred thousand rupees in three months. When I come back I can give my children plenty of food and have a happy family." She thinks about this in silence for a while and adds, "But then Adnan won't love me because other men will have looked at me and he'll be jealous."

She's caught in a dilemma: she doesn't know whether to keep the little security she has by staying with the erratic and unloving Adnan or strike out on her own and reenter the Heera Mandi bazaar. Both alternatives are doomed.

Adnan has arrived at Maha's house and some of the family are getting a decent meal of *roti*, chicken, and salad. Adnan and Maha sit on the mattress eating with Sofiya and Mutazar. Maha's children by other fathers wait in the other room, peeping around the door. There are eating hierarchies in many Pakistani families, and the poorer the family the more strictly these are enforced because there's less food to go around. Senior males are given priority and those lowest in the pecking order eat last: they have what's left when everyone else has finished. Females—especially children—are low in this hierarchy and a woman's children from a previous liaison are as good as invisible.

Adnan leaves and Nisha sits next to me on the bed bowed over her meal and picking at bits. She shreds her *roti*, makes it into a little pile, and pushes the chicken around the plate. She has no appetite at all. She walks, sunken-chested, into the bathroom, and I turn to Maha.

"What's the matter with her?"

Maha shakes her head. "TB, I think, for the past three or four months."

"Has she been to the doctor?"

Nisha shuffles back in and Maha motions me to be quiet. She whispers, "One visit to the doctor costs two hundred rupees and then there are more visits and more. What do I do: give her medicine or food? Adnan won't help."

"Mutazar went to the hospital just a few weeks ago," I comment. We all went with him when he had some stitches removed from a wound on his finger.

"Mutazar is Adnan's son and Nisha is someone else's child."

Nisha's pale blue polyester *kameez* is hanging off her shoulder blades. Her limbs are bent and she can't straighten one of her arms.

She pulls up her sleeve to show us a swollen, deformed elbow. I don't understand how Adnan can refuse to pay for her treatment or how her mother could have ignored her condition for so long. But Maha is preoccupied with her husband; her concern for their relationship is her whole life and her main, sometimes only, topic of conversation. The children she had with Adnan are valued because they tie him to her. The children she had with other men do not enhance the marriage: they serve only to complicate it because the children's food has to be bought with Adnan's money.

Maha claims that Adnan doesn't care much about his stepchildren. He says they are destined to be *kanjri* in the brothels of Heera Mandi. I think he's being optimistic. The only place that Nisha seems destined for is the graveyard.

maha has two children who no longer live with her. I discovered their existence when we were looking at some old family photos. I can't tell whether Maha's earlier silence on the matter is because she doesn't care or cares too much. She had four children with her *sayeed* husband: two sons in addition to Nisha and Nena. When he abandoned her, Maha's husband took the boys and left her with the girls. The sons had some value, whereas the daughters were going to be prohibitively expensive to marry off: they would have to be given an enormous dowry to compensate for the shame of their origins, and even then, no decent man would consider them.

Maha's family has such complex dynamics. The two youngest children—Adnan's children—have the lion's share of love and attention. They are fed and washed and treated with far greater care than their half-sisters. Mutazar, the son, is especially spoilt and frequently indulged. The two older girls receive poorer treatment and their clothes are nowhere near as new or as clean.

There is yet another child: Ariba. She's 11. When I first met the family I thought she was a servant or an impoverished friend of one of the girls because she looks very different from her two older sisters. Nisha and Nena are fair-skinned and fine-featured; Nena possesses the large, lustrous almond eyes of classical Indian beauties. Ariba, in contrast, has dark skin—considered ugly and very low class in this society—and her clothes are ragged and far too big. Her mother brushes her three other daughters' hair but she rarely brushes Ariba's.

At lunch today Ariba stood on the periphery looking as if she was in the wrong place. Perhaps she was: there was no plate for her. No one had told her to go away, but she was lingering, clearly excluded. She found a space on the corner of the mattress and ate bits of *naan* that Mutazar, her little brother, threw to her. There was virtually no meat left and no salad.

Ariba is learning how to be tough and streetwise. She spends a lot of time outside the house. This is unusual for a girl of 11 in Heera Mandi, but Maha doesn't seem to care about the dangers she faces. Ariba annoys me but I am also desperately sorry for her. She tries to steal from me whenever she can, but she's an inept pickpocket: I can feel her scratching around in my bag for money. Sometimes she throws a towel or a blanket on the floor between us so that she can scrabble underneath it to find my purse. When I turn to look at her in the midst of this furtive searching, she smiles fulsomely but nervously. She was successful once: she took seven hundred rupees ($12). I thought of telling her mother, but I knew she would get a brutal beating and her status as the pariah of the family would be confirmed. I was sad she did this, because if she had asked me, I would have gladly given her the money. But Ariba would never have asked. She would never have assumed that she could be given anything. She gets nothing without a fight.

After days of persuasion and negotiation Nisha is finally going to the doctor's. The visit is being treated like a family outing, and everyone squeezes into a two-seater rickshaw for the journey. The clinic isn't far from the house. It caters to the women of Heera Mandi, and to the poor of the inner city.

The consultation process is confusing. Dr. Qazi, a dour, high-speed medic, holds his surgery in the middle of an ailing throng. On the right-hand side of his desk are the women, most of whom are hidden behind a curtain. On the left-hand side are the men. Patients stand or sit in a queue that moves closer and closer to him so that, in the last stages, they sit next to his desk and can listen to, and participate in, the consultation offered to the patients immediately before them. It costs thirty rupees (about fifty-eight cents), but for this you get to hear about everyone else's problems as well as being given advice about your own.

Many of the patients are in desperate need of a miracle. Most are malnourished. The doctor's office is filled with limp, fat-bellied babies with oversized heads. The women patients over the age of 20 are divided into two groups: the shockingly thin and the seriously obese.

A frail girl is carried in and lies coughing blood onto the floor of an anteroom. She's about 15 or 16. Her skin is like parchment pulled tight over slender bones. Her mother stands next to the bed holding the girl's tiny baby. Dr. Qazi pays a fleeting visit and says that nothing can be done. The doctor's office empties immediately and the patients reassemble as an impromptu and uninvited audience to watch the girl die.

Once the death has been observed to everyone's morbid satisfaction the spectators vanish and rejoin the line waiting to see Dr. Qazi. Word goes around that it was tuberculosis. No one is surprised. Tuberculosis is one of the biggest killers in the developing world, and in places like Shahi Mohalla, it is approaching an epidemic. It's highly contagious, and a large proportion of the Pakistani population is infected, but it only wreaks havoc on those whose bodies are already weakened. It is, overwhelmingly, a disease of poverty, poor nutrition, and unhealthy, cramped homes.

Nisha is petrified—she thinks she's going to share the same fate as the coughing girl. Something else is upsetting her too. Perhaps the sight of the blood-stained floor has stirred memories of her childhood. She's sitting in Dr. Qazi's line, holding my hand and telling me about her father. "He hit my mother all the time. She was so frightened. He took her jewelery and gave it to another woman in Heera Mandi. My father punched her and kicked her in the stomach, and there was blood everywhere and the baby died."

Maha had been five months into the pregnancy when the kicking induced her miscarriage. Her husband had grown tired of her himself, and even though he'd decided to profit from pimping her out, he still became angry when the clients made her pregnant. After some practice, he developed such a talent for do-it-yourself abortion that perhaps Maha was relieved when he abandoned them and they had to find their own way home to the comparative safety of Heera Mandi.

Nisha is trembling as we take her for her x-ray, but she manages to forget her fears in the excitement. Maha and the children want to enjoy the entertainment. "It's expensive," Maha exclaims as all the family form a scrum around the X-ray machine to get maximum value for money.

"She has advanced tuberculosis of the joints." Dr. Qazi's diagnosis is swift and he doesn't feel the need to explain anything about her condition or the danger of contagion. He hands Maha a list of medicines and tells her to come back at some unspecified date. "She will recover," he tells me in English, "but she has to take the medicine and to eat well and rest. The most important thing," he stresses, "is that the parents take time to care for her." This last form of treatment is probably the most unrealistic prescription he could offer.

The Tawaif

Places like Heera Mandi are not new, and dancing and sex have been linked on the Indian subcontinent for millennia. For centuries, women like Maha have lived by selling their beauty, youth, and skills. Maha is from the Kanjar, one of the region's prostitute groups. Her mother, her grandmother, and her great-grandmother were all in the same profession: part of the generations of women who were born, raised, and trained to please men.

Pakistan's culture is a hybrid, a fusion of two great civilizations that converged in northern India: the ancient civilization of the Hindus and the newer civilization of the Muslim invaders, who ruled over large parts of India from the thirteenth to the eighteenth century. What we see today in Lahore are the remnants of Islamic and Hindu social practices that produced the *tawaif*—the courtesan.

Three thousand years ago religious prostitution flourished in Hindu temples throughout the subcontinent. Pubescent girls were married to gods and dedicated to temples where they performed ritual dances. The temples provided land grants to support the performers, but many women also had to supplement their income by selling sex. In time their daughters, too, were dedicated to the deity and so the cycle was perpetuated. This is not an archaic practice: its legacy continues even today in the *devadasi* tradition of India.

The Hindu caste system ranked people according to their occupation and ritual purity. Hundreds of subcastes—of traders, artisans, warriors, and priests—were arranged in an inflexible hierarchy. Social status was determined by birth: the son of a potter became a potter, the children of sweepers became sweepers, and the daughters of dancers and prostitutes inevitably followed their mothers into prostitution. Many of India's entertainers—the singers, dancers, minstrels, and bards of the subcontinent—were from the lower castes and often associated with prostitution. Women who performed in public were the antithesis of respectable Indian women, idealized as secluded wives and daughters. But although they were born into these disreputable castes and destined to be entertainers from the moment of their birth, they didn't prostitute themselves indiscriminately: their families were often retained by the aristocracy and sex was only part of the service they provided to their patrons.

Dedicating girls to temples is now very rare in northern India, where hundreds of years of Muslim rule destabilized the administration of Hindu temples and, in some cases, led to their destruction. But the system of Hindu prostitute castes has continued to flourish even in Muslim areas. Islam does not endorse a caste system—in fact, it promotes the equality of all men—but when Islam expanded into South Asia, it adapted to the Indian social environment and absorbed many of the basic principles of the caste system. Some of the Hindu prostitutes converted to the religion of the Muslim conquerors, but even so, they remained very near the bottom of a complex system of social stratification. Today, descendants of the Kanjar—a Muslim entertainer and prostitute group with obscure and vilified origins—live and work in Lahore's Heera Mandi.

Kathak

Maha dances slowly at first, concentrating on her footwork, her feet striking flat on the floor and her ankle bells chinking in a slow, rhythmic pattern. She smiles, lifts an eyebrow, and swings her hair so it falls in a silky curtain over her face. The music changes tempo, the pace of the *tablas* accelerating, and Maha motions her daughters to turn up the volume of the tape deck. She's dancing energetically now, her feet moving faster, her arms held high. Even though it's the middle of the night, it's still oppressively hot in her cramped rooms and, as she dances, the fabric of her green *shalwaar kameez* adheres to her back. Her face and neck are beaded with sweat, tendrils of hair sticking to her skin. She is breathing fast and excitedly—transformed. Maha loves dancing: it has been her life, her passion, and part of her livelihood since she was a child.

Orthodox Islam forbids singing and dancing on the grounds that it may lead to a loss of self-control and then to debauchery and fornication. The Mughals, the Muslim rulers who controlled large parts of India between the sixteenth and eighteenth centuries, did not see entertainment this way. Dancing and singing were considered to be forms of refined culture, and patronage of the arts was a symbol of Mughal status. The emperors employed thousands of artists and took *kathak*, a dance-theater form long associated with religious themes, out of the Hindu temples, changing its emphasis so that it became the favorite dance of the Muslim imperial courts. When Maha dances in her room in Heera Mandi, she performs a debased but still recognizable version of *kathak*.

Kathak is sublimely elegant, apparently effortless, and rigorously, punishingly difficult. The accomplished *kathak* dancer's feet, hands, and eyes must be tightly coordinated. Her hands have to be trained to twist, first the right hand and then the left—palms up, palms down—as the dancer steps forward and backward, then side to side, her head and eye movements flawlessly synchronized. The turning of a circle must be deft and seamless with the correct number of steps and an exacting, faultless motion of the arms. Her fingers must be perfectly and delicately controlled. It is a highly developed skill that is won only with long and patient practice. No one in Heera Mandi is trained in *kathak* dancing today. It takes years to master; it is expensive to employ a dance teacher; and audiences no longer understand the complexities of a dance form that is supposed to tell a lengthy story. Maha had only a limited training in *kathak*, and her daughters have had none. The older Kanjar women say it is sad; they say you have to go to Delhi to learn *kathak* properly.

Some of the essence of *kathak* has been incorporated into the world of Indian films, mixing it with a less refined, more accessible form of dancing and a good measure of Bollywood glitz so that it appeals to a popular audience. It is the same in Heera Mandi. Maha's dancing is elegant, but also lewd by Pakistani standards; her hand movements are sensuous, she kisses her fingers, lifting them into the air, stamping her feet, grinding her hips, and smiling suggestively. She doesn't have the discipline of the classical performers but she is still a proud heir to a precious tradition.

Anarkali—*Pomegranate Blossom*

The Mughals originated in central Asia, drawing their customs from an Islamic world in which women were secluded chattels and rulers maintained enormous *zenana* (female quarters) in their courts and palaces. The emperor had absolute control over his wives and concubines. Singers and dancers performed exclusively for the royal household, and beautiful dancing girls became concubines of the emperor and lived for the rest of their lives in vast hareems. Emperor Akbar kept five thousand women in his hareem and Emperor Arungzeb is said to have kept even more.

The Punjab Civil Secretariat in Lahore is a strange building that has had many incarnations. It was originally a tomb built in the early seventeenth century, then it became a residence, and for a time it was a British church. Today it is an archive housing books, city records, a few old maps, and, tucked into a corner, a white marble sarcophagus engraved in intricate detail. A little notice declares it to be the tomb of Anarkali—Pomegranate Blossom—the nickname given to Nadira Begam, who was a favorite dancing girl in the hareem of Emperor Akbar. Legend says that Anarkali fell in love with the emperor's son, Prince Salim, and that while the emperor glanced in a mirror he caught a glimpse of the lovers' longing gaze. In a fit of jealous rage, Akbar ordered Anarkali to be buried alive. The heartbroken prince never forgot Anarkali, and when his father died and Salim became Emperor Jahangir, he had a marble sarcophagus made for her. On one side of the sarcophagus are engraved the words "The profoundly enamoured Salim, son of Akbar," together with a Persian verse, which declares:

> *Ah! Could I behold the face of my beloved*
> *Once more, I would give thanks unto*
> *God until the day of resurrection.*

The story has become a Lahori legend, but one that is in dispute. An eminent historian of Lahore claims it was a fabrication dreamed up by an English merchant who visited the city in the early seventeenth century. The sarcophagus, he maintains, belonged to one of Jahangir's wives, not his murdered lover. But it was not uncommon for straying women to suffer a fate similar to Anarkali's live entombment. When the British envoy Sir Thomas Roe met the great Mughal Jahangir in 1616, he mentioned in his diary that a woman of the hareem had been caught with a eunuch, one of the castrated males who were allowed access to the emperor's women. The eunuch was cut to ribbons and the unfaithful woman was buried in earth up to her armpits and left in the sun, moaning about the pain in her head until, eventually, she died. Anarkali may be a myth, and the sarcophagus that is said to belong to her may lie forgotten in the dim and dusty secretariat, but her legend lives on in the name of Anarkali Bazaar, the busy main market of Lahore.

Less exalted aristocracy followed the example set by the Mughal emperors: they patronized the arts and kept mistresses both for pleasure and as a measure of social status. Only the richest men could afford the attentions of the most expensive courtesans. *Tawaifs*, though, offered far more to Indian men than a claim to status: they provided the romance and companionship that men could not find in their arranged marriages. The *tawaif* was the South Asian equivalent of the Japanese *geisha*. In the 1820s, Abbé J. A. Dubois, a French missionary, wrote that in a country where courtesans abound, they "are the only women in India who enjoy the privilege of learning to read, to dance and to sing. A well-bred and respectable woman would for this reason blush to acquire any of these accomplishments." Sixty years later K. Raghunathji wrote a book about prostitutes in Bombay in which he claimed that the elite Muslim dancing girl "is generally ready of wit, is more cultivated than a married woman, and owes much of her fascination to the fact that in a country where wives are not fit for society, she is a most charming and pleasant companion."

Chaklas —red-light areas—prospered in traditional and colonial India by providing two sought-after services: they supplied beautiful women for sex and witty company, and they were also the scene of *mujra*, traditional singing and dancing performed for the delight of a rich and cultured clientele. The salons of the best *tawaifs* were respectable establishments where the sons of the nobility, gentry, and intelligentsia were sent for an education in classical music, Urdu poetry, the language of love, and sophisticated etiquette. Some served as salons where composers, musicians, and writers sought their inspiration. To the outsider, seduced by the image of urbane glamor, it was a world of sensuous and exquisite delights. The *tawaifs* in these establishments were the elite of the subcontinent's prostitutes, and they became the lovers of powerful men. They always formed a minority of prostituted women, however, and very few ever became wealthy and respected outside the *chakla*. Below them the hierarchy of prostitutes descended in steep steps marked by a woman's beauty and upbringing. At the very bottom of the ladder were desperate women who sold sex in order to scrape together the meanest existence. Heera Mandi is what is left of a traditional red-light area and, like the *chaklas* of the past, it has its great *tawaifs* — its successful courtesans — but it is home to an even greater number of poor women who have no other way to survive.

The Village Family

It's a momentous day for the family living next to Maha. Their first-floor home is crammed with excited visitors and the place is a shambles. Jumbled piles of clothes lie on broken *charpoys;* a precarious tower of boxes is packed with a dust-coated muddle of items; and the threadbare rugs are ingrained with bits of debris and polished to a shine by grime and wear. The large family — a mother, a father, seven children, and a few grandchildren — all live in two rooms. The older daughters have children of their own: one has a toddler and an emaciated baby whose sallow skin hangs in crepelike folds. Maha raised her eyebrows when I told her I was going to visit them. "What? They're the village family," she scoffed. Everyone in the courtyard repeats the same thing: the family came from a poor village last year and don't know the ways of the city. The women can't dance and they can't sing. They can't even speak properly. They're nothing but low-class folk from the jungle.

The father of the village family is sitting in state among the anarchy. Two of the sisters bubble with anticipation and explain that they're going to Dubai on an airplane. A promoter from Heera Mandi is arranging the trip. It'll be the first time they have been out of the country—although one of the sisters interrupts and says that she's been to Karachi twice. This Dubai trip is a major career break for them and they're thrilled. They will go to Dubai as dancers, but the real money will come from selling extra services.

A large, tattered suitcase is dragged out and a fat man arrives in a rickshaw. He's brought a brand-new case in tan-colored plastic that both sisters claim as their own. Fancy *shalwaar kameez* are pulled out of unlikely storage spaces under the mattresses and from behind the sofa and are folded untidily and stuffed into the cases. Moments later, one of the brothers pulls them all out again. The clothes are vivid: lilac, purple, pink, red, and green. They are made from clumsily stitched synthetic fabrics and some are finished with embroidery and sequins. The women are so proud of their finery. They ask me to inspect the *ghungaroo*, the bells that they will fasten to their ankles, and the jewelry they will wear. I tell them their things are lovely—and that they themselves are beautiful—and with their radiant smiles and their infectious excitement, they truly are.

At three o'clock the suitcases are heaved down the stairs and into a waiting taxi. The Dubai-bound women climb in with one of their brothers. They wave up to their mother who is leaning out of the window, weeping and blowing her nose on her *dupatta*. A large group of people assemble in the courtyard to watch the farewell. "See you in three months," the women call.

Another of the courtyard's families is monitoring their departure closely. The people in the household living directly opposite my landlord, Iqbal, are also new arrivals in the *mohalla* and, like the village family, they are not yet accepted by the long-term residents. The father is an old, bone-thin, reformed drug addict. He has a son and a couple of daughters, and at least one of the girls is in the business. The rooms in his house are big and very sparsely furnished. A hundred years ago it must have been a very imposing residence, but it has long since slid into decay.

I've been introduced to a confusing succession of people in this house. Some of them are relatives of the family and some are girls who have been brought in from villages to work as prostitutes. The father manages them and acts as their pimp, and everyone is presented as some kind of relative. One ugly young girl, who always wears the same red dress, has a suggestion of a beard and no breasts. She is a boy.

The father is usually very busy. He scouts for custom in the street or sits on a *charpoy* in the courtyard waiting for business. He chats and jokes with passing men, and they must like what he says because they accompany him to the house for more laughs.

When I visited on a hot afternoon in May he wanted to talk about his dead wife and to show me her photograph. He seemed genuinely upset. We drank tea and he took me to the window and pointed into the courtyard at the beginnings of a little shrine covered in what appeared to be bathroom tiles and decked with Shia Muslim flags and *panje*. "He's a very religious man," someone explained. "He's building that shrine with his own money."

A one-room building stands next to the emerging Shia shrine. It has a low, corrugated-iron door and no windows. Inside a couple of *charpoys* balance on stilts of uneven bricks to lift them out of the monsoon floods and thwart climbing rats. The room is the pimps' den and the place from which they monitor the local women. The drug pushers congregate there too, waiting for customers who want to buy hashish, heroin, or pharmaceutical drugs. Mushtaq is one of the most important men: he's big and dark and handsome and spends his time relaxing, loose-limbed, on a tatty *charpoy* or slowly patroling the courtyard and *galis*. Traditional families of prostitutes did not need men like Mushtaq and his friends to pimp for them: they had their own clientele that they had built up over years. The new women arriving in Heera Mandi do not have these networks and they cannot solicit for customers—most do not even go out of their rooms—so instead they rely on the pimps to bring in the clients. Some of these men aren't just intermediaries: they operate their own businesses, importing women from the outside and putting them into the trade, whether they are willing to work or not.

They're a worrying-looking bunch: big strong men with black
moustaches and thin ones with gray stubble and quick eyes.

Nautch *Girls*

Today the links between dance, art, sex, and power are loose, but
these connections continue to draw on long South Asian traditions.
The Mughal Empire waned at the beginning of the eighteenth century and official support for the performing arts declined. Throughout the nineteenth century, dance became ever more clearly associated
with prostitution. Classical dancing was still the preserve of elite
courtesans, and lower-status entertainer castes continued to perform folk dances for less sophisticated audiences, but the gulf that
separated these women and the types of services they rendered nar-
rowed dramatically.

Tawaifs and entertainer castes lost many of their traditional patrons when the British removed a large part of the native elite in
northern India. By the time the Punjab was annexed by the British
in 1849 the weakened Mughal rulers had already been usurped.
About fifty years earlier, Lahore had been captured by the great
Sikh leader, Ranjit Singh, who allegedly had a retinue of 150 dancing girls, and for a time the city became the center of a short-lived
Sikh state. When the Sikhs were ousted by the British, the demand
for *tawaifs* and the service industries that catered to the old rulers
collapsed. The British deputy commissioner wrote in the 1868 district census report that there had been a relative decline in the population of Lahore's Walled City. "Since annexation, a class, which at
one time formed a considerable proportion of the population, has
been gradually dying out and its ranks are but scantily recruited . . .
the class of retainers, courtiers and hangers-on about the late Lahore *darbar* [court]."

In the early days of the British raj, the colonialists combined local
practices with the privileges of conquest and took Indian women as
concubines. They did not, though, become patrons of elite courtesans

and sponsors of the arts. The old ways lived on in princely states that maintained a degree of independence from the British. In places like Lucknow, local Muslim rulers actively encouraged the maintenance and development of traditional culture. Things were different in areas under direct British control. The new rulers rarely understood the culture, the etiquette, or the refinements of highly polished, Persianized Urdu. Most did not even enjoy the dancing of the women whom they called "*nautch* girls"—a corruption of *nachna,* the Urdu verb "to dance." Traditional *kathak* dancing is not an erotic spectacle, even by prudish Victorian standards. The *Punjab Gazeteer* of 1883 stated: "Dancing is generally performed by hired *nach* girls and need not be further mentioned here than to say that it is a very uninteresting and inanimate spectacle to European eyes." The dances can take hours and incorporate much technicality and symbolism that is entirely lost on those who are not versed in the art.

This indifference to the "*nautch*" was not universal among the British, especially as the nineteenth century drew to a close. After the Indian mutiny of 1857 great efforts were made to underline the superiority of the British by encouraging social distance between the white rulers and the natives. Taking local women as concubines became a subject of disapproval and then of scandal. British women were imported into India as respectable wives to replace Indian mistresses. This distancing coincided with the Victorian social purity movement that spread from England to the colonies. English and Indian prostitutes suffered increasing stigma and, by the end of the century, a powerful "anti-*nautch*" movement drew support from both British colonialists and some sections of Indian society. Despite the decorum and modesty of their performances, even the most highly trained and sophisticated dancing girls were reviled as lewd because everyone knew their origins.

Courtesans and ordinary prostitutes had been tolerated, and sometimes encouraged, in precolonial India, but under the British, the distinctions between prostitutes were obliterated and they were all lumped together as criminals to be policed. They still did not constitute a single community, however, because they remained divided by religion into Hindu, Muslim, and Christian prostitutes and into gradations according to wealth and training. Among the prostitutes of the cities of north and central India it was the descendants of the courtesans, the singers and dancers of the old feudal courts, who formed the elite of the red-light areas.

Tourists at the Sufi Shrine

The anniversary of the death, or *urs*, of Data Ganj Bakhsh Hajveri is a festival in Heera Mandi. Data Ganj Bakhsh lived in the eleventh century and he is Pakistan's most important Sufi. Sufism is the mystical branch of Islam, first spread into South Asia by wandering Sufis. Data Dabar, Data Ganj Bakhsh's mausoleum, is the spiritual center of the Pakistani Punjab; it lies just on the other side of Bhati Gate, a fifteen-minute walk from Heera Mandi. It's always busy—especially on Thursdays, when it is packed with devotees—and, during the *urs* commemorations, our part of the city becomes crammed with pilgrims.

Devotees have flocked to the city throughout the day. Some arrive in specially hired coaches; others come by train or on public buses. A field by the side of the Badshahi Masjid has been transformed into a giant encampment. Enormous tents cover a third of the ground and food is being prepared in dozens of metal cauldrons called *degs* over charcoal fires. The tents, the musical entertainment, and much of the food is paid for by important feudal families from the rural areas: it is a form of paternalistic benevolence that they grant to their tenants and other poor folk. In Fort Road one of the local drug dealers has opened a water stall. An enormous vat sits underneath an awning and the water is offered to sweaty, dehydrated pilgrims in communal plastic mugs.

Groups of women circulate among the hundreds of men on the field, sauntering about and sitting on the grass to sing for a few rupees. Their real income comes from selling sex: the singing is just a form of advertising. These women aren't from Heera Mandi: they are women from the villages, and they service their clients in the

cheapest lodgings available. Despite all this competition from outsiders, the local prostitutes are still preparing themselves for some
robust business. Most of the tourists are poor villagers, so the
lower-priced women are likely to be in greatest demand.

Some of the devotees are giving me a really difficult time. They
block my path in the street and force me to walk in piles of rubbish.
They make unfriendly comments, like "Get out of Pakistan" and
"*Kanjri,*" and a few throw stones. A couple of old men became angry
because my veil slipped for a time and part of my hair was uncovered.
A broken brick was lobbed at my back for no apparent reason. Rural
Pakistanis' attitudes to women are even more conservative than those
of city dwellers. A foreign woman on the streets—even a veiled one—
is an affront to many Pakistani men who come to Heera Mandi.

I sit by the side of the road under the shade of a tree, watching
the tourists going to visit the fort and the Badshahi Masjid. Most
are men, but a few women accompany their fathers and husbands,
dressed in their best clothes, hobbling by in shiny, rarely worn
shoes, proud of their finery and nervous of the outing. They give me
puzzled sideways glances and very few smiles.

A middle-aged man stops in front of me and grins. He's unusually unattractive with a filthy *kameez* and one closed eye. "I love
you," he says.

"No, you don't. Not really," I reply.

He speaks to me again in English. "Yes, I do. I very love you.
I very, very love you."

I gather my shopping and prepare to flee.

"Look." He pulls out a wad of notes from his pocket. "I love you.
I have much money. Come with me. I very love you. How much do
you want?"

There's no time to reply because I race back to the house, but
even then he doesn't give up. He hangs around outside for hours,
patrolling up and down the road, keeping his one good eye on the
building. Perhaps he thinks I'm playing hard to get.

The streets, especially near Roshnai Gate, are filled with attrac-
tions. A man performs a show with a wretched dog and a couple of

old monkeys that he pokes with sticks. Another popular attraction is a ten-foot-long homemade cinema constructed from metal sheets held together with giant rivets. It's a weird contraption, shaped like a rocket lying in a wheeled cradle. Customers peep at a small screen through tiny holes running along its side while a young man stands at the tip of the rocket turning a spool of film in front of a crude projector. Another man plays a soundtrack that has absolutely no connection with the pictures being shown inside. It's a desperately amateurish affair, but it has lots of patrons. The owner of the mobile cinema invites me over to have a look. Among the visions on offer are knife fights, a man being impaled on a stake, another being decapitated. As a special treat, it is promised that sex scenes will be sandwiched between the next killings.

By ten o'clock at night no women are visible on the festival ground. Hundreds of men are sleeping on the grass around the darkened perimeter of the field. Some areas have been set aside for bathing: a hosepipe supplies the water, and part of the grass has been transformed into a big, muddy pond in which a dozen half-dressed men lie cooling themselves. Outside the tents masseurs are giving treatments and people congregate in little groups to talk and eat. Some of the tents are reserved for sleeping, but the very largest are being used for entertainment. A highly accomplished *qawwalli* singer performs songs in the largest tent. These are songs that are a form of worship in the Sufi tradition, a conversation with God building slowly, layer upon layer, to a melodic, stirring crescendo that creates such ecstasy in the listener that they are able to draw nearer to Allah. All of the *mohalla* is witness to this devotion: the sound of the singer's voice carrying from the Hazoori Gardens right up into the heart of Heera Mandi bazaar.

The New Nawabs

Heera Mandi has been in decline for decades and elite prostitution has been leaving the *mohalla* for over fifty years. In his memoirs of colonial Lahore, Pran Neville wrote of the joys of courtesans and dancing girls, and of Heera Mandi, "which came to life at night with its reverberating sounds and glittering sights when fun-loving Lahoris would flock to it for entertainment."

In those days, the most sought-after patrons were the Punjabi landed gentry and the urbanized, intellectual Lahori elites. This historical memory remains alive for some of the women who continue to work in Heera Mandi. When Maha speaks of one of her former clients, a rich landowner, she calls him a *nawab* and links him with what she believes to be a golden age when women like her were respected *tawaifs*.

Heera Mandi sparkled less and less as the rich began to leave their *havelis* in the inner city for the conveniences of new homes in spacious suburbs. Heera Mandi's patrons also changed. After Independence in 1947 Pakistani entrepreneurs began to develop modern industries. Together with men from the expanding middle-class bureaucracy, these industrialists and businessmen provided Lahore's prostitutes with a new type of patron. Some of the women left the *kothas* of Heera Mandi to entertain their clients in other parts of the city, and this process accelerated when the military government of Ayub Khan, pursing a policy of stricter Islamization, closed the red-light area in the 1960s. Deprived of their traditional places to live and work, many of the women of Heera Mandi moved into the world beyond the walls. A public outcry about the dissemination of vice led to the return of at least some of the women, but they were officially restricted to singing and dancing, and even that had to take place at carefully circumscribed times.

The traditional Pakistani elite stopped visiting the *mohalla* twenty or thirty years ago. The men who today are cabinet ministers, diplomats, bureaucrats, and senior army officers tell me that they and their friends came to Heera Mandi in their youth but ceased visiting long ago. They say it was not just because they became older and wiser but because the diversions of Heera Mandi became illegal and, most importantly, because the place became unfashionable for the rich and influential. The only really powerful men who visit today are the godfathers of criminal fraternities who hold meetings in the *kothas* of the *mohalla* in the quiet hours just before dawn.

Twenty years ago the bazaar was filled with function rooms. Many of these have closed now because customers demand much less dancing these days and rather more basic sexual servicing. Some of the old establishments that were on prime sites on the main road from Tarranum Chowk to Taxali Gate have been replaced by shoe shops. Only a few of the remaining ones keep professional musicians, who provide the dancers with live musical accompaniment. Instead of performing to the sound of a harmonium, a *tabla*, and a *dholak*, many women now play tapes and CDs on a "deck"—a tape and compact disc player. These "deck functions" are more popular among the clientele because they are so much cheaper.

Today's rich have little desire to demonstrate their social status by supporting a beautiful courtesan who speaks flawless Persianized Urdu and sings impeccable *ghazals*. Instead, they wear Rolex watches and drive Land Cruisers. Poor men don't want to see singing and dancing either—they can see plenty of it on the television or videotapes. Expensive prostitution rarely happens in the *mohalla* and houses in pleasant suburbs are just as likely to be brothels as the ghettos of Heera Mandi. Rich men might not come here anymore, but the girls of Shahi Mohalla continue to visit high-status clients in hotels or discreet brothels in select suburbs. The men who patronize Heera Mandi are lower- and middle-class men—those without the sophistication or the finances to afford the new elite prostitution scene in plush hotels with girls who speak English and have designer handbags. The world has moved on from the days of landed elites and cultured *nachne walli*. The *mujra*—the singing and dancing display executed with skill by the courtesans of the past—is almost dead. The market for entertainment has changed and Heera Mandi's women are being dragged along—and down—with it.

"In Those Days It Was Different."

Two types of sex workers live in modern Heera Mandi: the traditional Kanjar families and the new entrants to the profession. Most of the Kanjar say that they came from India during Partition. When the British left in 1947, the subcontinent was divided between Hindu India and Muslim Pakistan. Many of those caught on the wrong side of the divide—the Hindus and Sikhs in Pakistan and the Muslims in India—fled their homes. In the process Lahore was transformed from a multifaith city to the Muslim city of today. Punjab was the scene of horrendous bloodshed during Partition, and the older Kanjar women of Heera Mandi report that they escaped from the communal warfare of north Indian cities with nothing but their jewelery and the clothes they wore. Most claim an illustrious but unverifiable past as favorites of *nawabs* and maharajas, and for years after Partition they were known in Heera Mandi by the Indian city or region from which they had migrated.

Today only a handful of these extended families can be identified. They have been swamped by recent recruits to the trade, and the divide between the two types of prostitutes—the established families and the new arrivals—is increasingly vague. In Heera Mandi very few people are certain about exactly who is a "real" Kanjar because their origins in pre-Partition India are so vague. It is only the women of the traditional Kanjar families who cling fast to the old distinctions.

On a hot June morning I sit having breakfast with a group of elderly Kanjar ladies. Most are in their seventies or eighties and spend their days drinking tea and eating mild, sweet *paan*. They talk about the past and how much better it was before things were spoilt by cheap women in their low-class *khoti khanas*— the name given to the worst kind of brothel run by male pimps.

"When we were young we lived in India. That's where we are from," the oldest woman explains. "In India we sang for the aristocracy. We worked for the Maharaja of Patiala. We were his servants.

"In those days it was different. We had high status and were respected. We were trained as singers and dancers and we had to practice for hours every day. We began when we were about seven, and then we started performing when we were fourteen or fifteen. The classical singers had the highest status and the good musicians had high status too. Only the lowest kind of entertainers went to bed with the men. People looked down on them, but we were respected and other women would listen to us sing. It wasn't just the men who came to see us."

Maybe things weren't really quite as rosy as this, but that is how the older women like to remember their youth. They were artists, they insist. Their skills had nothing to do with sex—except, perhaps, sometimes. Their singing careers lasted about fifteen years and they retired when they were 30.

"When we came to Lahore in 1947 there was no one here. Many of the houses were burnt and the people who lived here—the Hindus—had gone to India. We came here because this was the diamond market: we knew it was where the singers and dancers lived.

"Coming to Lahore was difficult because we had to leave everything behind in India. We carried our gold and jewelry and we lived for a while by selling what we had. When it ran out, my elder sister's daughter began dancing. She danced in some movies and sang in the theater and in big stage programs, but it was hard because she was the only one supporting our entire family. In the end the other girls went into the business too.

"Until twenty-five years ago lots of people came to Heera Mandi to listen to the music—women as well as men. They would come to big functions and sit around the singer while she stood in the middle of the room and sang. It was good in those days, but all that has changed. Nobody bothers with singing and dancing anymore. We were trained for years, but today nobody does that."

The old ladies are chewing and sucking *paan*, their mouths stained red because they eat it continuously. They have a specially worked silver *paandan*, a box filled with little pots of spices, pastes, betel nut, and tobaccos. A silver plate sits on the bed, its pile of betel leaves kept fresh under a wet cloth. It's a simple and addictive pleasure and they pay careful attention to detail, adjusting the combination of spices to make the *paan* just the way they like it.

A younger woman—in her late forties or early fifties—adds her own memories to the story. She talks about how the standards of the area have gone down: how the area has been tarnished and their reputations have been spoiled by the nature of the business in today's Heera Mandi.

As I'm about to leave she asks me angrily, "Why do the men come here and then leave us with babies and never come back? Why do they do that?"

I say it happens everywhere, not just in Heera Mandi, but she doesn't agree. She thinks there's something especially bad about this *mohalla*.

"Don't they care about their children?"

I say that the men have come here for pleasure and that, after the pleasure, they will forget all about the place and the women they've loved. I add that they probably realize that their children will be cared for.

"It's true," the old ladies agree. They smile and nod and chew their *paan* while they watch the young women of the family playing with their babies. They've had very long lifetimes in which to get used to the idea.

I ask if they'll ever leave Heera Mandi. "Why should we?" they say indignantly. The younger woman adds that she has been here, in this house, all her life. She doesn't know anywhere else. "We have nothing to be ashamed of. This is our home and we're respectable people. What other people think is their problem."

Women like these old Kanjar ladies are the heads of their households: there are no real husbands in their families and men are short-term guests. In a complete reversal of most Pakistani families, women hold power among the Kanjar. Women earn the money and women manage the profession. They are known as *naikas:* they decide when a girl is ready for the business; they decide on her clients; they manage the courtship; and they decide how to spend the money that's earned.

The Kanjar look down on the new entrants to the business because they say they are cheap women who have no artistic standards. Some of these women have migrated to Lahore from impoverished rural areas; some have been sold to brothels by their families; others are married to men who pimp them out; others flee abuse at home and have nowhere else to go and no other way to earn a living. Most find themselves in brothels run by men who pocket the profits from their labor. Some are locked into a system of sexual slavery. I am told that in the poorest parts of Heera Mandi some girls are held in chains after they tried to escape from their pimps. I've never seen these girls. No one ever does except the clients. Perhaps they don't exist—but the very thought of them is sufficient to keep young brothel workers compliant and in thrall to their pimps. Unlike the Kanjar women, the new arrivals in Heera Mandi do not have family networks to help them or a rich heritage to give them psychological support.

The traditional, intensely patriarchal culture of the subcontinent exploited Kanjar women, but this older system of prostitution provided an element of protection that is missing in the newer, more vicious structures of the Lahori sex trade. Old women have a place in Kanjar society because they run the little family firms that make up a large part of the sex trade. But now there is a new breed of manager: a professional class of pimps, agents, and procurers who are rendering the managers of the established industry redundant. Some of these people are from Heera Mandi, some from outside, and they are increasingly wealthy and powerful. The future for elderly Kanjar women is ever more fragile: the modern-day executives want workers in their teens and twenties—women who will be discarded as unemployable shortly after they reach thirty.

"I Like Them to Be like Girls."

Tarranum Chowk is permanently and frighteningly busy. In the middle of a hot June day I spot a young *khusra* dodging the rickshaws as she skips by the cinema. She's throwing a flimsy *dupatta* around her shoulders. Her hair is plastered with henna and scraped on top of her head so that it looks like a little pile of dung. Her beard is just beginning to show, and her eye makeup has melted slightly in the heat.

She hesitates and stops to look at me. After a few moments she reaches out, adjusts my *dupatta,* and traces her fingers over my face. "Beautiful," she laughs and kisses her fingertips.

She catches hold of my hand and tells me her name is Tasneem. We dance back down toward the Badshahi Masjid and turn into a narrow passage barely lit by the sun filtering through the space between the houses forty or fifty feet above us. The walls of the buildings are a dark, drab gray, and there are so many flies feasting on the rubbish that the mounds of rotting food seem to move.

Tasneem takes me into a house leading off the alley. Twelve or thirteen *khusras* are lying on cushions, all dressed in pretty suits. They're friendly and seem pleased to see me: it gives them a new subject to discuss. They make me sit in the middle of the room so that they can look at my hair and nails.

Tasneem sits next to me busily arranging her *kameez* and hoisting up her bra to make it appear as if she has breasts. Her performance starts a competition over who has the most womanly figure. On the surface it's good-natured banter, but there's a constant undercurrent of competition amongst the *khusras.* They're obsessed with who is the most beautiful and the most feminine.

One elderly *khusra* insists on showing me her breasts. They look very real. She says that she had also been born with a penis but that it had been removed. She begins to take off her *shalwaar* to prove the point, but I tell her that there's no need: she's clearly an authentic *khusra*.

Tasneem puts her hands on the floor next to mine. She has a man's hands, far bigger than mine, with long fingernails that are varnished in maroon and knuckles that are heavily creased despite the fact that she's in her early twenties.

"Wrinkles! Wrinkles!" the other *khusras* shriek. Tasneem flies out of the room in tears and hides behind the door with her face in her *dupatta*. She makes an elaborate display of being wounded and the others laugh even more.

Tasneem beckons me to follow her, and we climb up the spiral staircase right to the top of the building. She shares a room here with other *khusras*. Large, professionally produced photographs of them cover the walls. All the dancing girls—and boys—of Heera Mandi have these framed glamour portraits in a place of honor, often adorned with gold tinsel. In the photographs, heavily made up and wearing their finest clothes, they become stars who have escaped the dismal poverty of the *mohalla*. It's a fantasy but they love it. I'm obliged to note the beauty of the subjects, their obvious star quality, and the magnificence of their outfits.

As a rule Pakistani men are handsome in a very masculine way, and the *khusras* are no exception. The pictures capture butch men plastered in makeup, jewelry, and flamboyant clothes. Tasneem and her friends hold their prized portraits before me. They're so earnest that I can't say anything other than, "Very lovely."

Tasneem announces that I need beauty advice. I'm alarmed. She makes me sit on a little stool by the window so that she can see me in the light. She opens her makeup box, rummages through ancient, grubby cosmetics, and sorts out the most appropriate colors. She varnishes my nails in the same maroon color as her own, paints my lips the deepest burgundy, and makes my eyebrows two thick black lines. There's unanimous agreement that I look like a stunning dancing girl, but I'm shuddering at the thought that I have to pass through throngs of men in Heera Mandi looking like one of Cinderella's spectacularly ugly sisters. We walk home holding hands.

Tasneem has left her *dupatta* at the *khusra* house and so, with stylized sauciness, she mimics what any self-respecting woman would do in public—she covers her imaginary breasts with the end of my own shawl.

asneem likes being sent to buy cold drinks at the shop by Roshnai Gate. Sometimes I meet her as I go to buy food at the Al Faisal Hotel. She enjoys the opportunity to camp it up as she walks down the road, flirting outrageously with the men who make *roti* and the customers drinking tea in the restaurant. They think she's funny and they laugh, making comments to each other. She responds by pouting, flicking her hair over her shoulder, and making a convoluted performance of retrieving money from her bra. A real woman—even an experienced prostitute—would never dare to behave in such an overtly sexual manner.

Today, some new people are sitting in Tasneem's room when we return with Cokes. They're men who look like men. One is slim with fine features and the other is short and square with a luxuriant moustache. They tell me that they're not like Tasneem and the others; they don't live in Heera Mandi, but they visit every week or so. The slender one comes to dress up, dance, and sell his services. The other is a customer. Both men are married and they want to know if there are men like them in England.

I tell them that men who have sex with men in my own country don't have to dress up as women unless they prefer to. The *khusras* think this is interesting, but they're skeptical. The man with the lush moustache says that it's not a nice idea. "I like them to be like girls," he adds. Tasneem looks serious. "The customers want us to be pretty. No pretty dress, no pretty face—no money."

The *khusras* take it in turns to dance. Tasneem is teaching me a combination of foot movements and the slender visitor performs an energetic routine of twists and flourishes. And then the dance group's guru shows us how it really should be done. About 40, with a muscular build, heavy features, and long black hair, she's the striking person I saw in the street on one of my first nights in the bazaar. Her name is White Flower and she owns the *khusra* house. The others whisper and nod to me knowingly. "She's the best," they explain, and they are right. She dances like a woman, with coquettish glances and swaying hips. The graceful movements are perfect even if her body shape is not. Her audience is enthralled and so am I.

"I Was Born This Way."

The *khusras* of Heera Mandi are similar to the *hijras* of India. Both are often described as "half man, half woman." Most of the *khusras* I know in Heera Mandi were born biologically male: they look like men and they have a penis and testicles. A few—very few—are genuine hermaphrodites. Some of the biological males undergo surgery, often paid for by their clients, to remove their sexual organs. In the process, they rise a notch above their still complete friends. Their superiority is based on their lack of a penis and on the more feminized appearance that they begin to develop once their bodies are deprived of the testosterone produced by their testicles.

Ambiguous sexuality has an acknowledged tradition in Hinduism: deities in Hindu mythology have male and female essences, some transforming from one gender to another. The idealized *hijra* is an ascetic linked with a goddess, and with fertility, and has a recognized power to bless or to damn. It is customary for *hijras* to dance at houses where a baby boy has been born or where a marriage is taking place. If the *hijras* are paid well, they leave placing a blessing, but if they are poorly rewarded, they vent their anger and curse the newborn child or the marriage bed. These spiritual powers inspire a mixture of dread and ridicule even in modern India. It's the same in Pakistan: in Lahore, the *khusras* dance, unsolicited, at weddings, births, and circumcisions, and they rarely leave without a respectable profit.

Islam considers homosexuality to be an abomination, but the Muslim world has a history of third genders and eunuchs playing important roles in society and in the courts of Muslim rulers. A few eunuchs were born with ambiguous genitals, but the vast majority were castrated either as children or as young men. They were valuable as loyal guards, teachers, and administrators, or as slaves who could be used for sexual recreation by adult men. Male prostitution has long flourished in the cities of the subcontinent, and today about 10 percent of the dancing girls of Heera Mandi are *khusras*.

Many men have sex with men in Pakistan; they also have sex with boys. Homosexuality is derided in public, but it is accepted, providing it remains a secret. The men involved in homosexual acts don't perceive themselves to be homosexual, and the men's families won't perceive them to be homosexual either. Most Pakistani men marry and produce children: their extramarital sexual preferences are irrelevant as long as they can maintain a respectable public face. Having sex with other men or boys is not associated with stigma providing a man takes a dominant role in sexual encounters. It may even reinforce a man's masculinity and status because he is sexually dominating others. It is the receptive partner who is despised and ridiculed. He is labeled as submissive and passive—like a woman—and *khusras* and boys fall on the feminine side of the gender divide. Most of the same-sex relationships that are found in Heera Mandi are profoundly unequal: encounters between men and boys, or between men and *khusras*, and they faithfully mirror the power imbalances in relationships between men and women.

The *khusras* imitate and exaggerate women's mannerisms: talking in high-pitched voices, fluttering their hands, swaying their hips, and looking coquettish. They grow and paint their fingernails; wear thick, painstakingly applied makeup; and pay attention to their hair, which must be as long and as glossy as possible. Whenever the *khusras* want to impress me with their attractiveness, they undo their hair, arrange it over their shoulders, and tilt their heads back to emphasize its length. One of the most serious and cruel punishments that the *khusras* can inflict upon one of their own is to cut her hair.

Khusra behavior is a caricature of feminine conduct that extracts the most useful things from the women's world and leaves out the rest. Most importantly, they don't observe the restrictions placed upon women's conduct. Dancing in public would be unthinkable for ordinary Pakistani women, and even the most devout *khusras* in Heera Mandi fail to observe that most important sign of Pakistani Muslim femininity: they never observe *purdah* and they wear veils only as a prop for some extra-piquant flirting.

Heera Mandi's *khusras* are organized into houses based around a guru, a teacher or leader. Each guru has *chelas*, disciples or apprentices who live with them and are trained in obedience to the community's rules. In theory it's like a large family. Each person must do as the guru instructs and not act independently. The *mohalla* has a hierarchy of gurus, and the top guru commands a respect that verges on reverence. A network of *khusra* houses throughout Pakistan allows young *khusras* to move from one house to another with astonishing speed. A *khusra* who is expelled from one community will not be able to find support and succor in another. This is supposed to reinforce solidarity among *khusras*, but not all gurus are kind: in Heera Mandi many operate as brothel managers, squeezing money, work, and spirit out of the youngest and most vulnerable in their care.

It's difficult to discover the origins of anyone in the *mohalla*, and it's especially difficult to know about the early life of the *khusras* because most say the same thing: "I was born this way." A few elaborate and say that they came to Heera Mandi because they could not be themselves in their villages—that there was no space for them, they felt wrong and out of place, and they shamed their parents. Some were forced out of their homes, and others left of their own accord, interpreting their journey to the city and to this life as something good—an escape. Others say they were sexually abused as children and many were prostituted. Indeed, there may be even more boy prostitutes than girl prostitutes in Pakistan. They work in garages, as helpers on buses, as assistants to truck drivers, or as waiters in tea shops. They are everywhere: impoverished boys between the ages of 9 and 14 who sell sex for a pittance. Tasneem says she was raped when she was young—so many times she cannot remember—and that it made her the way she is: she cannot be a man and she will never be a woman. She can only be a *khusra* and the only way she can live and survive is by selling sex. Men do not want her now her body is hard and she is able to grow a beard, but she can still cross the gender divide if she wears a pretty *shalwaar kameez* and matching lipstick.

A Prostitute with Honor

(Monsoon: August 2000)

Today is August 14, Pakistan's Independence Day, and crowds of boys and youths are running wild in the streets, lighting firecrackers, shouting "Long live Pakistan," and waving cricket bats. It's not polite for a guest in Pakistan to be anything other than enthusiastic about this special day. I've been asked a dozen times, "Which is the best country: India or Pakistan?" I dissemble. I say both countries are nice and, in the bakery, I don't join in the men's curses against what they think of as the bullying, imperialist, giant neighbor but only tut diplomatically at the evil that is India.

The street is busy with men walking to Iqbal Park on the far side of the Badshahi Masjid, and from the roof terrace I can hear the proceedings of a political rally relayed by loudspeaker. The importance of the day and the number of visitors to the area are matters of concern to the local authorities. The garbage Dumpster that usually stands surrounded by mounds of putrefying garbage at the top of Fort Road has been removed, and the men's open-air and very public toilet that runs along the adjacent wall of the old water tank has been scraped to within a few inches of cleanliness.

The Hazoori Gardens are full of men and the Pakistani flag is everywhere: on posters and embroidered or printed on badges and caps. Some of the youths are wearing Western-style clothes and T-shirts printed with the words "I love Pakistan." Visitors squeeze their way through the entrance to the fort, but the main action is taking place fifty meters in front of the Alamgiri Gate. A large crowd has gathered and a policeman is shouting and wielding a *lathi,* a long bamboo cane. At the center of the crush two young female Japanese tourists are drinking cans of Coke. They're wearing tight tops and spray-on jeans. Even in Shahi Mohalla the men rarely get to see a woman in anything as revealing: it's like an open-air sex show. The girls move off toward the *masjid* and the all-male audience moves with them as a gigantic, fused body.

A flourishing trade is being conducted by Roshnai Gate. *Tangas* loaded with four-foot-long shafts of ice race by, leaving a watery trail up the road. The ice will be chipped into blocks and sold to those without refrigerators. Street vendors have fashioned lavish displays of fruit and vegetables on their carts. The mountains of mangoes are the most impressive. South Asian mangoes are peerless—sweet and highly fragrant—and in July and August they are everywhere in Lahore: assembled in tall towers at juice stands, piled on carts, on stalls, and in baskets strung over bicycles. It's worth enduring the monsoon to feast on mangoes. Their skins and big flat stones, sucked bare of their orange flesh, lie scattered on every road.

My landlord, Iqbal, has been working on a new painting over the holiday. A woman comes each day and sits in a modern yellow rickshaw outside the house as Iqbal arranges his easel and works for thirty or forty intense minutes while the light is perfect. He is worried about the bright colors on the rickshaw—worried that all we will see are the colors and not the woman inside. I watch as his frustration and concentration change the shape of his face.

He needs a break from this place, a way to find a different perspective, but I think this will never happen. And if it does, Iqbal might be able to find a kind of peace, but there will be no more pictures that have a voice: when Iqbal Hussain paints, he does so from a troubled heart filled with sadness and an anger that he cannot speak.

Ama-Jee

I'm pleased to be back in Heera Mandi despite the damp, enervating heat and a worryingly unhygienic interruption to the water supply. I have a crate of mangoes in my room, and I'll take the best to share with Maha's children. I'm always happy for the first few days of my visits: I catch up with all that I've missed; I eat plenty of hot, tasty food; and I wander around the *galis*. I sit in the vaulted calm of the Badshahi Masjid. I settle into my room and call on my friends. It's like returning to a sorely missed home—one that hides a disturbing secret. I love this place and hate it in equal measure. It's both fascinating and loathsome. When I'm away from the *mohalla* I am desperate to return, yet when I'm here, I can't wait to escape.

Maha has moved yet again during the two months that I've been in England. She's left the "American kitchen" and come home to the courtyard where she's taken four rooms on the second floor of a decaying mansion opposite my own room. Little else has changed. Nisha is still ill. She's thinner than ever and her bones are becoming progressively more deformed. She refuses her medicine and sulks when she's forced to swallow the tablets. She says they're too big and they taste awful, so most of the time no one bothers to monitor her medication and the tablets sit in the cupboard gathering dust.

Adnan is still the same indifferent or absent husband. He told Maha that she can go to the Gulf to earn some money, so she's been working hard: her weight has stabilized and she's busy transcribing page after page of lyrics in her songbooks. She practices for hours to extend and perfect her repertoire, and harmonium and *tabla* players visit the house every day to run through the routines.

Leaving the children is going to be difficult. Maha will be away for three months. Not only that: she's leaving them in the brothel quarter at a time when three of her daughters are ripe for the business. Her family refuses to help unless she can pay them more money than she will earn. They may even put the girls into the business themselves.

Maha is resourceful and she has a plan. She's found a woman who will care for the children. The new home help lives in the house and works for little more than her food and a bed. She looks to be about 60, but she might be much younger. Whatever her age, she's quick-witted and fast on her feet. She has bright green eyes that sparkle in a dark, fine-boned face and her plaited hair is long and silvery white. They call her Ama-Jee, which means "respected mother." It's a good name for her.

Adnan seems unaware of Ama-Jee's presence. He doesn't care about anything except his drugs. When he visits Maha's home he no longer comes for sex but to smoke hashish and heroin. He changes out of his ordinary clothes and wraps a *dhoti* around his waist. He sits in a corner, bare-chested and cross-legged, often in the semi-darkness, and he lights his hashish and relaxes. I can see the tension ebb from his body. At times I pity him: he's not the cruel, vicious client of my imaginings—more a fraught and lackluster man struggling with addiction.

"I want Mumtaz to die," Maha cries. Tonight, yet again, Adnan smoked, gave her a few rupees, and then left. Maha had made a big effort to look attractive: she's wearing a tight black *shalwaar kameez,* a lot of shiny pink lipstick, and dramatic coatings of eyeliner. And still he wasn't tempted.

"I hope she gets cancer," she spits through tears. "It's black magic. She's using black magic."

Ama-Jee sighs. She's a servant, but that doesn't stop her giving Maha the benefit of her wisdom or loud reprimands whenever she feels that Maha is behaving badly. She has some good advice to hand out on the subject of Adnan, and I'm glad, because Maha is obsessed with him and his vilified wife. Whenever we meet she asks the same questions. "Does he love me? Does he love Mumtaz? Does he like sex with Mumtaz more than with me?" I have no idea of the answers.

Tonight we've reached a new low. Maha has decided that Adnan prefers his wife because she has larger breasts. I look at Maha's pneumatic bosom barely contained in a tight *kameez* and think that the possibility of finding a woman with larger breasts is remote. To clarify her point Maha pulls up her *kameez*. "Look. Are they nice? Do you think Adnan will like them?" She looks wild and tearful. I assure her that her breasts are absolutely wonderful and that they would be the envy of most women. I also add that it is unlikely that Adnan will measure his love by the size of her breasts, and if he does, she would be better off without him. Maha looks at me as if I'm deranged and without an ounce of understanding.

For once Ama-Jee agrees with Maha. She nods and speaks with an age-old wisdom and a wicked grin. "Men like big tits, big smiles, good hips, and a little *kusi*."

Shadi — *Wedding Ceremony*

Men and women are equal but different in the eyes of Islam: they have separate roles and live in segregated worlds, but both can enter heaven. Real life is not quite as generous to women. Men can have up to four wives, and although it's usually only better-off men who can afford to support more than one wife, husbands can initiate divorce with such astonishing speed that aging or disappointing women are very vulnerable. Wives don't have such prerogatives: they're permitted only one husband at a time, and although they can divorce erring spouses, they invite social death by doing so. A divorced woman in Pakistan is pitied. I'm viewed with compassion by many of the women here because I'm divorced. My friends in Heera

Mandi say I should look for a new husband quickly while I still have the youth and looks to capture a man. I don't think they're optimistic about my chances.

Sex outside marriage is technically illegal for everyone in Pakistan, but in practice this rule does not apply to men. A marked double standard operates. Women must be chaste daughters, faithful wives, and celibate widows—good women whose sexuality is under tight control. Men, in contrast, are at liberty to have extramarital affairs and indulge themselves with lovers, like those from Heera Mandi, who will be condemned for participating in those same relationships. In places like Shahi Mohalla, society has created a group of women, distinguished from the chaste daughters and faithful wives, who live under another form of male control: simultaneously celebrated for their sensual beauty and derided as unclean.

Men's first wives are usually selected for them by their parents: they are not romantic partnerships but arranged marriages allying two families rather than two individuals. For this reason it's unlikely that these marriages will end in divorce. Men who can afford the expense of another wife may enter second and subsequent marriages for more selfish motives: for passion or romance.

Romantic love is forbidden to most people throughout South Asia, and yet, paradoxically, it is also extolled as a cultural ideal in film, music, and literature. Men visit Heera Mandi for sex, but some also come in search of the love and companionship they don't find at home, and some of the relationships that they form with women of the *mohalla* are intense, happy, and enduring. A far greater percentage of women, though, will have short-lived marriages: a few may have a *shadi*, a wedding ceremony in which they wear a beautiful gold embroidered wedding dress, but even fewer possess a *nikah nama*, a legal contract of marriage. Most will talk about a *shadi* as a way of speaking about the transactions involved in selling sex. A wedding in these instances can last a single night.

No woman in Heera Mandi claims on first or even second acquaintance to be a prostitute. They say they are *nachne walli*, dancing girls, and those who are involved in longer-term relationships say that they are married. They have to say this because, otherwise, it means that they are committing *zina*, unlawful sex.

Young girls aren't encouraged to commit themselves to a single patron. Their teenage years are their prime years, the years in which the forward-thinking dancing girl must maximize her earnings from clients who are willing to pay a premium for youth. Tying herself to one patron only makes sense if he is extremely wealthy and can offer the teenager the resources usually provided by a number of clients. Once a woman is in her twenties, demand for her services begins to drop. It is at this point that she will consider settling down with one man: she will stop taking other clients and she won't dance in the bazaar.

Even the most infatuated client rarely stays with his Heera Mandi "wife" for long. His visits will tail off after a couple of years and then cease altogether. The woman will have to look around for a new patron, but by then, she will be a little older and a little fatter. Her options will be more limited and her clients will have less money and status. Maha's first long-term relationship was with a rich high-status *sayeed*. Now she's with Adnan—far less wealthy and a drug addict to boot. She knows that when he leaves, as he almost certainly will, her chances of finding a decent husband will be almost zero.

The most successful of Heera Mandi's courtesans form relationships with powerful Pakistani men. Farida has a sumptuous home in Gulberg, a rich suburb, that she shares with her sisters and their children. The house stands in its own compound with carefully tended lawns. Guards rush to swing open the gates as we arrive, servants appear, and a couple of gardeners glance up from pruning the bushes and watering the flowers with the fine spray from a hosepipe. We're taken into a tiled porch bordered with terracotta planters and trailing jasmine, and then through an elaborately carved wooden door into a spacious and shady hall where the air-conditioning gives merciful relief from the summer heat.

I wait with Iqbal in a room full of finely worked wall hangings, sumptuous curtains, and sofas scattered with unyieldingly hard cushions until Farida floats in, petite and vivacious. She is one of three daughters from a family in Heera Mandi who was trained in classical singing and dancing. By the time she was 15, she had become an accomplished performer. Luckily for her family she was also pretty and adept at pleasing her patrons. Her sisters didn't possess her charm and beauty, but they could still command a sizable fee, and in time the whole family moved away from the *mohalla* to the comfort and respectability of the suburbs.

Farida became the mistress of a well-known Pakistani politician. There were other women in his world—his wife and the other pretty girls from Heera Mandi who passed through his life, rarely staying more than a few weeks—but for years he came to Farida for relaxation. It was his favorite way to escape the rigors of public life and the pressures of his official family. After almost a decade, the patron found a new love and no longer visited. Today, like so many of the women of Heera Mandi, Farida is alone. Unlike most, though, she possesses some special advantages—a house, a savings account, and, locked in a bank vault, a safe-deposit box brimming with jewels: gold bracelets from Dubai; ruby and emerald sets; and other tokens of love from her former patron that she now sells to sustain her lifestyle.

Ursula is another of those rare women who have escaped the brothel, and her prim manner rarely betrays her upbringing in the *galis* of Heera Mandi. I refuse a gin and tonic in her reception room because it's not respectable to drink and Ursula never touches the stuff. She's unlike any other woman I know in Heera Mandi: she doesn't carry the *mohalla* in her every gesture.

Ursula's husband is twenty years older than her. They have a *mut'a* marriage, something that until a couple of decades ago was common among the Kanjar. *Mut'a* is a temporary-contract marriage, the length of which can be specified in advance: it can last hours, days, months, or years. During the marriage the husband supports his wife financially, and as her part of the bargain, the wife provides sexual, domestic, and emotional labor. When the contract ends, the obligations end, although a man is required to support any children born as a result of the marriage. Ursula has had a succession of year-long *mut'a* marriages, and when one marriage comes to an end, her husband decides whether to get married all over again: she's on a kind of rolling contract.

It has paid her well: she has a lovely house in the exclusive suburb of Defence. Ursula's home, like Farida's, possesses manicured lawns, an army of servants, thick carpets, and uncomfortable sofas. Her husband arrives from the Polo Club in his new Mercedes, striding in, exuding confidence, and giving us a hearty greeting. His English is impeccable, the result of a British education, and he sits like a potentate in a large leather armchair conversing upon many topics of international interest. He gives no indication that he is embarrassed by being caught at the home of a former Heera Mandi dancing girl. It's acceptable for rich men to have many wives, and extramarital affairs are almost compulsory for a man of status.

Respectable Wives

I have lots of noisy neighbors. The loudest open their shutters and play their tape decks at full blast. Sometimes, several families set themselves up in competition. Punjabi songs are the most popular with their catchy melodies and rhythmic beat, but they have to battle with the sounds of rickshaws, youths on motorbikes, and the shouts of children playing or fighting. When I sit on the rooftop and concentrate I can also hear shouting from the house twenty yards down the street, and over the weeks it has become part of the background noise, blending with the other commotion.

I've never seen the young woman who is being shouted at. I know of her existence only because I've been told that she's there and I can hear her being instructed and reprimanded by others in the household.

The house is owned by Rani. She's a friendly woman, and I spend the odd evening with her, sitting on the floor and eating spicy fried potatoes. Rani's eldest son is 22, and it is his wife who is the household's ghostly figure, hovering somewhere upstairs. She's a "proper" wife: she is not, and never has been, in the sex business. The sons of *nachne walli* don't marry dancing girls: their first wives are girls of unquestioned virtue. Many come from outside the *mohalla*, from respectable but impoverished families in the villages.

In Pakistani society control over female sexuality is a marker of a man's status and honor. To win respect men must compete with, and dominate, other men in the public world, and they must keep tight control of the sexuality of the women in their family. The sons of the *mohalla* are unable and unwilling to exercise this kind of rigorous control over their mothers, sisters, and daughters, but they can do so over their secluded and protected wives. In a bizarre twist, the daughters of these women will enter the sex trade when they are teenagers.

In this brothel community, women have a high status because they sustain their families with the sale of their bodies: they have a clear financial value. Ironically, respectable wives like Rani's daughter-in-law are only one step removed from domestic slaves. They do all the household chores: they wash, clean, cook, sew, produce children, and care for them. These women don't get paid, and so are often treated with contempt. For Rani's daughter-in-law, the benefits of being a genuine wife may not be all that clear. Unfortunately, I cannot ask her directly. She is completely secluded, and I will never see her.

Tibbi Gali

Tibbi Gali is the cheap end of the Lahori sex market. It's easy to get lost in this part of Heera Mandi, and I've been told repeatedly not to come here alone because of the danger. It's supposed to be a terrible place, with bad women and many thieves. Absolutely everyone says the same thing—even the people who live here. The women say it of themselves: they are *kharab*—bad, spoiled.

A group of young men sitting in the bazaar insist that the women in Tibbi Gali are the very worst kind of women. "They can't dance or sing. They sell their bodies, that's all," one says indignantly.

"And they do it for nothing—for one hundred and fifty rupees [$3]," says another, disgusted by the contemptible price.

"The old ones are even cheaper," a youth tells the others. "An old woman costs twenty rupees [thirty-four cents]." They shake their heads and laugh. It's the price of a bottle of Coke.

There are several ways to enter Tibbi Gali, but today I walk down the road from the bazaar to Taxali Gate, past the shoe shops, left down a narrow road, and then left again up into the *gali*. It seems like a long walk. The *gali* climbs and winds through tightly packed buildings, alleys branching off and tapering into narrow, gloomy passageways. Women lean against door frames, lie on rope beds or blankets in tiny, dark rooms flanking the *gali* and passages. A few younger girls saunter by in brightly colored *shalwaar kameez*, chatting to their friends, their body movements supplely relaxed and loaded with provocative sexuality. Respectable women don't move like this; they keep their bodies stiff and strictly regulated.

The road turns again and then curves quickly downhill to where Tibbi Gali joins the road to Bhati Gate. At several places along the *gali* the bricks and paving stones covering the drains have collapsed, leaving gaping holes in which wastewater pools with sewage and decomposing rubbish. At one point, not far from the shoe market, the collapse is so severe that almost the entire ten-foot-wide *gali* has fallen into the drains. A couple of stepping stones allow the constant stream of pedestrians to do a deft hop over the filth-choked subsidence. The most vacant of the matted-haired drug addicts stumble through the morass oblivious. Passage along the *gali* is made difficult by other kinds of obstacles. There are goats tethered to the houses, piles of fodder, disoriented drug addicts, peddlers selling fruit and vegetables, and testosterone-fired youths riding motorbikes at insane speeds.

I can see fifty or sixty women. Even in the shadow of their little rooms you can see the brightness of their lipstick. Only rarely does it appear as if it's been applied with the aid of a mirror. They don't look like women in the other parts of Heera Mandi: these women are clearly for sale. Most are looking at me quizzically. Some are stony-faced. Others shout to their friends to come and see the *goree*, white woman, passing through their world. They peep out of the darkness and laugh. Some smile and reply to my greeting. They are friendly, these women who are the most despised of the despised.

The Luxury of Purdah

Most women in Pakistan wear a *dupatta,* a piece of cloth about three feet wide and seven feet long. It's used to cover the hair and breasts—those signs of female sexuality considered to incite men to lust. Wearing a veil is a sign of respectability, but no one wears a *dupatta* in Tibbi Gali: there's no point. These are the lowest order of women, and they are granted no comforting pretenses when they work here. Pretensions only cost them money, so they remove their *dupattas* to show that they're public women for sale.

Many women wear a *chador,* which is like a very large *dupatta.* Some wear a *burqa,* a long black cloak that covers them from head to foot. *Dupattas, chadors,* and *burqas* are part of observing *purdah,* which means "curtain," and is the real or symbolic divide between the sexes. Islam is very positive about sex, providing it takes place within marriage. The problem is that attraction between men and women is considered so inevitable and so uncontrollable that it will lead to chaos if it isn't regulated. Separating the sexes is essential to prevent fornication and maintain social order. Men and women have to live in separate worlds. Ideally, they are divided by walls and physical space and women are secluded in their homes. Well-off families can afford to keep their women in the luxury of *purdah* as a sign of status, but the poor don't have the finances to cocoon females within their family, so when these women appear in public, they are divided from men by the use of a veil. These rules on veiling don't apply to Westernized, educated elites, and throughout Lahore rich Westernized women wear a *dupatta* as a hair accessory rather than as a head cover. Some walk around bare-headed, their veils no more than a sliver of gauze tossed over their shoulders. The old women of Heera Mandi, steeped in tradition, think such styles are brazen.

I spend a lot of my time walking, veiled, in the streets; I want to observe the *mohalla,* and occasionally I like to get out from between four walls. The local women ask me why I do this. They say, "Why do you go into the streets like a shameless woman?" We don't understand each other. Perhaps it's surprising that they still want to be my friend. I've thought about this often. I used to think it was because I'm foreign and am assumed to be rich and generous, but the reality is a little more complicated. Sex workers in all parts of the world are friendly: being nice to strangers is their trade. In Heera Mandi, women who stay at home most of the time enjoy receiving visitors because it breaks the monotony of the day. And, perhaps most importantly, having a foreign acquaintance is associated with status and with an international lifestyle. I've never had the slightest difficulty making contact with the women of the *mohalla:* higher-class women lean out of their windows and beckon me in, and in Tibbi Gali, they rush out of their rooms and take me by the hand. Most of the women are unfamiliar with the names "England" or "Britain," but they've heard of London, America, and New York, and they like to introduce me to everyone they know as their friend from one of these exciting places.

I may be socially valuable, but I'm also alarmingly indiscreet because I don't always adhere to the strict rules over the separation of the sexes. Sometimes I speak to men in the courtyard, and in the evenings, I sit with Iqbal on his roof terrace. Excuses are made for me, but I realize that I cause confusion. The idea that men and women can be friends—and no more than friends—is inconceivable, especially in the *mohalla.* I have a friend, a Pakistani man whom I've known for years. We meet occasionally when he flies into Lahore and we go for dinner at the Pearl Continental Hotel. Maha is always excited when I'm about to see him, coaxing me to put on more makeup. A couple of nights ago I returned from a meal with him and Maha was expectant. She wanted to know the details.

"How much did he give you?" she asked.

I looked askance, and she tipped her head to one side.

"Twenty thousand rupees [$337]?" she questioned, trying to read my face.

"Thirty thousand [$505]?" she gasped. "You were that good?"

Shame

A competitive, often cutthroat hierarchy operates in Heera Mandi. Hundreds, perhaps thousands, of women sell sex in the *mohalla,* but even though they live in the same geographic space, they inhabit separate social worlds and do not interact except during religious events. An unofficial grading system divides women. The elite are "A" class: young and beautiful, with rich clients and a quality background that breeds good manners. Middle-ranking women fall into the "B" category: they're not so pretty—perhaps older or with coarser manners. The cheapest women, like those in Tibbi Gali, are "C" category. A privileged place in this order has to be worked at, and women like Maha behave in a way that is meant to underline their status. Slipping down the hierarchy is the nightmare of higher-class women. They know that aging will cause their foothold on the career ladder to falter and then to fail catastrophically. To forestall this, and to compete with the constant supply of fresh girls new to the trade, they perform an unending, complex, and nuanced piece of theater.

In Heera Mandi, a rich, secluded, and veiled woman has honor. She has superior patrons, and because she is protected in her home and has fewer clients, her fees are much higher. There's a finely judged code about exactly when and how far a woman should veil. When a male visitor calls at Maha's house, she adjusts her *dupatta* in a manner that reflects his social status in relation to her own. If the man is important and has high *izzat,* honor, she sits on the mattress while covering all of her hair, her breasts, and her legs with a *dupatta* or *chador*. She may also lower her gaze and speak quietly. If the man is less important, her *dupatta* is looser and her hair often tumbles out from under the material. If he's of low status, the *dupatta* barely sits on her head and she jokes and laughs loudly while lounging on the mattress. If the man is a workman or a servant, she doesn't even bother to veil and the *dupatta* lies crumpled on the floor.

Maha manages what is called her "shame" very carefully. Shame is behavior that is considered demeaning for a woman: it's associated with sexual activity of any sort, with contact with men, and with the failure to control and discipline the body. Dressing inappropriately, being too friendly with men, smoking, drinking, talking, and moving in a relaxed way—all these things are "shameless."

Men and women must maintain physical distance, and they mustn't look at one another either. Orthodox Muslims are trained to observe *"purdah* of the eyes." Looking at the opposite sex is like a form of adultery. The women of Heera Mandi are shameless in their flouting of this social code. In Tibbi Gali, the women look men in the eyes as equals. In the fancy *kothas* of the *mohalla* the practice is different—more sophisticated but just as daring. The women dance, hiding their faces behind their hands and peeping through their fingers. They keep their eyes fixed on their clients and then, for a carefully choreographed few moments, they drop their gaze to the floor and slowly, longingly, they look back at him. It works every time: the *tamash been*—customers—are transfixed.

Manipulation of the veil and management of the gaze is an art. Women flirt with a glance that lasts a fraction of a second, and they are provocative even while wearing a piece of material that's supposed to obscure their sexuality. They fold and refold the cloth, they flick it around and tease it, and then they smooth and drape it in suggestive ways. Paradoxically, a veil can heighten rather than lessen a woman's sexual power. A woman clad in a *burqa* walking with the utmost decorum through Heera Mandi can make every man's head turn with the tinkle of her ankle bracelets, the bright polish on her toenails, and the heavy, intoxicating scent of her perfume.

Shela's Shop

Shela has a shop on a corner in Tibbi Gali. It's more like a cupboard than a shop: it measures three feet by four feet and only a few dusty items are on display—cigarettes that she sells one at a time, soap, sweets, sachets of shampoo, noodles, a selection of cheap, artificial jewelry, and hair bands, combs, and clips.

Shela doesn't usually sit in her shop but on a step on the opposite corner of the *gali*. She's a traffic hazard. The *gali* narrows here to less than six feet wide and pedestrians are forced to negotiate a bend in the lane as well as Shela. Peddlers struggle to pass through, heaving the baskets of goods that are strapped to their bicycles around the shopkeeper. Shela sits in the shade and never flinches. She must have an important position in the *gali*'s hierarchy to receive such consideration.

Shela appears to be doing well despite her limited stock. She wears gold bangles and lots of dangling earrings from her numerous ear piercings. Like so many women—both inside and outside the brothel quarter—she wears her wealth: it's the safest way to keep it. She's around 45, with handsome features and the confident, sharp-eyed look of successful madams. She must trade in far more than the goods on show in her shop. Whenever I take a walk in Tibbi Gali, she calls me over to drink tea. I add to the congestion by sitting with her in the shade. A little stool is brought out for me and an audience congregates until they grow bored by the spectacle of the foreigner drinking tea. A handful of men inquire about my availability, but Shela sees them off with a raucous shout.

A group of young women work near Shela's shop. They lean against the walls chatting and waiting for customers. There are four or five regular women who are here all day, every day. One of the friendliest is Sabina. She's around 20 and has a deformed leg that causes her to limp. When she's propped up against the wall, though, you would never guess that she has a disability. She was born in Tibbi Gali and is the daughter of a prostitute. Her own daughter—a child of 5—will follow in her mother's and grandmother's footsteps. Sabina thinks I'm funny but I've not worked out why.

One of the women is standing on the periphery. She's not joining in the conversation but is looking around at the others nervously and without making eye contact. She's tense and doesn't slouch against the wall. She stands rigidly at attention and doesn't seem to know what to do with her arms, folding and refolding them over her breasts. Her skin is unusually dark—almost black—and her face hasn't yet lost its childish curves. She has rounded cheeks, beautiful features, and the half-formed body of an adolescent. She can't be more than 13. I've not managed to speak to her, and when I catch her eye, she gives a stiff little smile that doesn't extend beyond her lips.

This girl is beautiful enough to be in a higher-class brothel, but her skin marks her as cheap. Black skin is considered ugly and bad. When the women of Heera Mandi want to damn someone, they'll shake their heads and say, "She's very black," or "He's rich but he's very black." It's considered an affliction.

Being fair-skinned is thought to be good. Being white and freckly is not. I have red hair, white skin, and freckles on my arms. In the summer I get a smattering of them on my face too. People in Heera Mandi think I have a contagious disease. On three or four occasions mothers have motioned to their children to move away from me when the sleeves of my suit have worked up my arms. Most don't say anything: they just look worried. Others are more direct. They say, "What are those spots on your arms?"

Shela's attention is caught by a new customer: a *khusra* is peering into the shop. She's well over six feet tall and her brown flowered *kameez* hangs over broad and painfully thin shoulders. She stoops further to speak to Shela. Her face is pockmarked and she has a deep, velvety voice. She has neither youth nor looks, and she certainly has few friends around the shop. She buys a single cigarette and leaves. Shela jerks her head in the direction of the *khusra's* disappearing back, "Pathan. Ass-lover," she says and the other women snigger.

A young woman called Nazia works further up the *gali.* She wears the thickest makeup I've ever seen. It forms a pink-white mask on her brown skin, and I thought at first that it must have some medical purpose. On closer inspection it seems that her foundation is so thick and clogged because she never removes the previous day's makeup before applying a fresh coating. The layers have been building up for months.

Her mouth shares some of the same characteristics. It's hard to identify individual teeth because they're so heavily furred by plaque and encrusted with food. I feel sorry for this neglected girl, but then I'm disconcerted to see that she's looking at me with compassion. "What happened to your hair? It's terrible," she cries. She sits me on the step next to her fat, middle-aged brothel keeper and tries to get my curls under control. She sticks some pins into my hair and uses a fine-toothed comb to try to contain the rest. It's torture.

"It's all wrong," she concludes and gives up with a sigh. She takes me by the hand and leads me down the *gali* to Shela's shop. She searches around in a bag and pulls out a plastic fuchsia headband. Then she counts out a few rupees and gives them to Shela.

"Here," she says pushing back my hair and fixing the band in place. "That's better. Now you look like a pretty lady."

Dirty Girls

Nazia is always in this doorway. It's noon—almost a week later— and she's sitting in the shade to avoid the sun. I join her and her madam, a woman of 50 or 55 with vast thighs and those familiar, sharp, brothel keeper's eyes. She's friendly but doesn't give much away. She tolerates me sitting in the front of her brothel because it sends more clients Nazia's way and because she wants me to help her get a visa so she can start up a little business in London. I say it's difficult and prohibitively expensive, but she isn't convinced.

Nazia is the only woman working in the brothel. Her madam does the haggling with the customers. They walk up and down the *gali* glancing at the women and moving briskly as if they're in a great hurry. The ones who are interested in Nazia negotiate a price with her madam. If it's mutually agreeable, the men disappear into a back room with Nazia. I've been sitting with them for an hour, and in that time Nazia has entertained four young men, all in their twenties and early thirties. The turnaround time is fast—perhaps five or ten minutes. Then the customers leave—again in a hurry. As the young men stride away down the *gali* there's a lot of splashing from inside the room. Two minutes later Nazia is back on her seat. I glance around at the returning girl and try not to look surprised. Nazia's bright mask is undisturbed by whatever goes on in the back room.

The *tamash been* are a mixed bunch. Most are young men, laborers, and not by any means the poorest. They look fit, and more than a few are good-looking. A couple of the customers stand out from the rest. A devout-looking man in his sixties with a big beard and a small hat called a *topi* is approaching us. I assume he's a *maulvi*—a holy man—and I brace myself for the preaching. He does launch into a speech, but it's only to ask for Nazia's price to be reduced. He is, he claims, a poor but decent man and he can't afford the going rate. Nazia's madam tells him to clear off, and we watch him wander down the *gali* trying to arrive at a cut-rate deal with the other women, with equally fruitless results. Ten minutes later, he's working his way back up. "He's got no money," the brothel keeper says with a dismissive wave. "And he'll take a long time."

Another unusual client stops to take a look at us. He seems out of place: a middle-class man in a poor area. He's in his fifties, and is wearing a clean suit in starched and brilliant white. A cloth printed with multicolored flowers and fruits is bundled bizarrely around his head and tied so loosely that it flaps over his ears. The effect reminds me of Carmen Miranda and I'm sure he is wearing lipstick. I think he's an aging *khusra*, but then I change my mind: he's another client. He smiles sweetly and speaks to me in English. "You know what this place is, don't you? You know what these women do?"

"Of course," I reply and add, "Do you live here?"

He rolls his hands over and over and then brings them to rest in what looks like supplication. Licking his lips, pursing them, and then licking them again, he explains: "No, I don't live here, but I like

to come to look at all the pretty girls." He glances around at the
women framed in the doorways and adds with relish, "There are lots
and lots of pretty girls." Before he moves off down the *gali* his eyes
linger over my *shalwaar*, and I'm glad that they're crumpled and
edged with dirt around the ankles. On reflection the grime might
have contributed to my allure. The essence of the thrill: going to the
dirty place to meet the dirty girls.

Jamila

Jamila's house is tucked between a shoe shop and a tea shop.
There's no electricity, no running water, and the uneven floor is
made of bare earth. There's no roof either—only a tarpaulin and a
couple of tattered plastic sheets. The rest of Jamila's room is open to
the skies, and when it rains, water cascades through the footwide
gap between the plastic and the walls, turning the floor to mud. The
sun shines brightly after the storms, sending the temperature outside to over one hundred degrees. Inside it gets even hotter, and
anyone caught between the muddy floor and the plastic roof is
steamed into stupefaction. Staying too long in Jamila's house on
summer days is to invite a severe bout of dehydration.

Jamila is in her early sixties, and I met her a few years ago on my
very first visit to the *mohalla*. She was still in the business then, entertaining poor laborers for a few rupees. Now, though, her long career in prostitution is over. She begs on the streets and in the
Hazoori Gardens and does a bit of pimping for the women in Tibbi
Gali. She has spent fifty years working her way down the prostitution hierarchy and has, at last, reached rock bottom. Her current
husband, Mahmood, is her fifth.

"No, he's the sixth," Jamila says making a very slow mental calcu-
lation. Her first husband bought her from a dealer when she was only
a child, kidnapped from her home in India during the agony of Partition. She is not like the Kanjar, born into this trade. Husband number
one had a *tanga* business and kept Jamila for his own pleasure and

profit until he replaced her with younger girls and she was passed on
to her second husband, a man who had a tea shop and kept her in the
back room for the clients to enjoy after tea and kebabs.

Mahmood, her sixth husband is sitting on the bed with his leg
lifted off the ground. He pulls up his *shalwaar* to show me a terribly
swollen ankle and an infected wound. He has a congenital limp that
hinders his movements, and a year or so ago he was involved in a
road accident when he failed to move fast enough through Lahore's
insane traffic. The wound has never healed.

Mahmood is a little younger than his wife, and he must have a
genuine soft spot for her; otherwise, he wouldn't be living with her
now that she's no longer commercially viable. He used to act as her
pimp and would stand in the road quietly soliciting customers while
she waited in the dark of the house. The lack of electricity and the
poor visibility must have been something of a boon. Now, with his
leg injury and a serious limp, he can't stand for any length of time, so
perhaps it's fortunate for him that Jamila's business has faded away.

Jamila enjoys being at home despite the heat. Eight or nine cats
share the room; a couple are fully grown and the rest are bony kittens. They sleep under the *charpoy* or among the planks, wooden
boxes, tins, and rags that have been chucked against the back wall.

Jamila has no children. For women in South Asia this is a calam-
ity because children provide security in old age and endow women
with status as fully qualified females. In Heera Mandi a woman
without daughters has a bleak financial future and, perhaps ironically, a surprising number will find themselves a victim of this cruel
fate—particularly those girls who are not yet out of childhood but
who enter the business and entertain too many clients too often, and
for too long. They become infertile through injury, infection, and
botched abortions. Jamila doesn't have any children, but she has
her cats, and she treats them with a gentle affection. A plastic bowl
sits in the corner of her room filled with bits of offal and chunks of
ice. She dips her fingers in and sifts through to find a tasty morsel,
shredding it and placing it in neat piles on a plastic mat. Those that
are too ill to stand are fed by hand.

"This is my son," she says putting a kitten no more than five inches long into my lap. "And this is my daughter," she clucks, tickling the tummy of its mother.

Jamila is feeding these scrawny cats with food that is superior to her own.

Swimming in Dubai

Only one of the two sisters from the village family is at home. The other one has already returned to the Gulf because her first trip to Dubai was so successful. The house is full of people and a baby is dozing in a makeshift hammock. A blue sheet is tied to the frame of a *charpoy*, and the baby is hanging in it, a fraction of an inch from the floor. When his family pass by they give him a gentle nudge that sends him swinging to and fro. His grandmother hangs her foot over the *charpoy* and pushes him occasionally to keep the momentum going.

"Dubai was so good," the eldest daughter says with relish. "There was a swimming pool and I wore a swimming costume." She points to bits of her body to show just how high the legs of the costume were cut and just how scandalously low the neckline plunged. She sucks in her breath through barely opened lips, raises her eyebrows, and gives me a knowing smile. The costume sounds as if it would be entirely normal in a Western context, but in Pakistan, the exposure of so much flesh in a public place can only be equated with the hottest pornography.

The swimmer is leaving to pay a visit to an agent who organizes dancing tours to the Gulf and she promises to get some cold drinks on the way home. She puts a *burqa* on top of her *shalwaar kameez* and pulls the black veil over her hair. Her face is covered with a piece of fine black mesh. It's possible to see through this cloth in bright sunshine, but the world appears dark and obscure, especially where the material folds, and in the shade it's hard to see at all. She'll be completely blind if she wears it on her journey down the unlit spiral staircase. An ancient female friend of the family sits cross-legged on a chair and watches the *burqa* being arranged. "Why don't you just use a *dupatta?*" she asks.

"Because I'm a prostitute with honor," comes the sharp reply.

Law and Order

A couple of policemen are sauntering around in the courtyard, their rifles over their shoulders. There was a raid yesterday on an alcohol shop and distillery, and they must be checking to see that the place has shut down. Alcohol is illegal in Pakistan: it is forbidden by Islam, but you can still find a supply if you have the right contacts. There are little unofficial distilleries in the villages and in the back rooms of houses in the cities. They produce concoctions that are so toxic they can make you blind — or dead.

Christians can obtain a license to buy alcohol from official "permit rooms." There are two in Lahore. The whole process is grubby and difficult. It takes a long time to find the government building that issues the permits, and then you watch an official — and his dozen helpers — shuffle papers around a desk for a couple of hours. The official is a memorable man, attractive and unpolished, and he spits so often that the concrete floor of his office is iridescent with gobs of green phlegm.

The permit rooms where you use the license to buy the liquor are hidden in the basement or round the back of the big hotels. You creep in past gray shuttered doors and men look at you through a grill. It feels like you're doing something dirty. Sometimes I wonder why I bother: the spirits are so rough I get a thumping headache after two gin-and-tonics.

The policemen in the courtyard are looking into the liquor factory. There's no one there and all the bottles have been confiscated. Perhaps the police are having a party up in the police station. It's impossible to tell which policemen are corrupt, and in Heera Mandi, most of them are involved in the business in one way or another.

They take bribes and receive sexual favors, and many stop men in the street and force them to pay impromptu fines. They pick on men from out of town because they are unsophisticated and the easiest to intimidate. Whenever the police find themselves short of money, they launch a cleanup operation in the *mohalla:* they throw a group of women into the *thanna* and then demand extortionate payments for their release.

Police work in Heera Mandi is very lucrative. Rumor has it that officers pay big fat bribes to secure a stint at the Tibbi station. Many of the police I see in Shahi Mohalla don't even work here. They have jobs in other, less desirable areas and just pop into the *mohalla* to have a bit of fun and do some unofficial duty collecting fines and protection money.

Tariq and the Sweepers

A sweeper comes to clean my room every day. His name is Tariq and he's one of the nicest people I've met in Heera Mandi: a big, warm, smiling young man with a lot of beautiful teeth. He tries very hard to please me, regularly polishing the same mirror two or three times with a piece of scrunched-up newspaper. Tariq calls me "sister." This isn't because he's being overly friendly but because he is a Christian and thinks that, as I am from the West, I must be a Christian too.

The Roshnai Gate Presbyterian Church is in Heera Mandi Chowk and it would be difficult to identify if it wasn't pointed out. Even the cross is small and well above eye level. It's painted a deep, dirty red—the same color as the walls. On my first visit to Heera Mandi, Tariq took me to see the building. It was something he was very proud of, and as we stood outside, I made encouraging but not necessarily truthful comments about its charm.

The interior of the building is more impressive. A plaque on the wall tells us that it was founded in 1908 by the Reverend G. L. Thakur Das and the Reverend H. D. Griswold. It can't have changed much in almost a hundred years. It's stark and the pale blue and yellow walls are fading to gray. There's a small cross behind the altar, and three plastic Christmas trees, and the words "Merry Christmas" have been stuck on to the wall. Dozens of narrow, multicolored crepe-paper streamers criss-cross the ceiling and flutter in the breeze from six ancient ceiling fans.

Today, as always, the congregation is made up of Tariq's relatives — one giant family of friendly sweepers. They play a confusing type of musical chairs throughout the service—getting up, milling around, and changing pews. A new person comes to sit by me every five minutes, and each time I turn around there's a sea of smiling faces.

Tariq's uncle plays the organ and his brother plays a *dholak* with all the flavor and rhythm of the best musicians in the bazaar. Four other boys and young men form the choir. Everyone is dressed in his or her best clothes. Tariq's three tiny children are beautiful and immaculate, and although his uncle is wearing a strangely feminine, white-lace *kameez*, he does so with a great deal of reverence.

The congregation sways and claps, and the little girl next to me jiggles around the pew with excitement. She's full of energy and barely contained. Every time I glance at her she looks up with wide, dancing eyes and a dizzy smile. Her pretty face is downy and her upper lip is covered with the tiniest beads of perspiration. She's half-singing, half-shouting, and at the end of each hymn, she sighs with satisfaction and takes some deep breaths to prepare for her next musical eruption. She's lovely and fresh and so unlike the other girls in Heera Mandi, who are miniature women at the same age.

Most of Pakistan's Christians trace their ancestry to the *churha* — the untouchables of the subcontinent who were considered polluted, and polluting, because they did tasks that Hindus defined as ritually impure. They came into close contact with blood, death, and dirt as a result of their occupations as cleaners, butchers, skinners, and scavengers. This put them right at the bottom of the religious and social hierarchy—so low that they were treated as subhuman.

At the beginning of the twentieth century missionaries in the

Punjab converted many untouchables to Christianity. The *churha*, though, were still unable to escape the stigma of their untouchability, and they continued to do the same kinds of defiling tasks. During Partition in 1947, the Christian sweepers of Lahore stayed on to become part of the new Muslim state of Pakistan. Islam is not supposed to have a caste system, but here in Pakistan, one exists unofficially. Tariq thinks there are around two thousand sweepers in the city. He says all of them are Christian and none of them are shown any respect.

"In the tea shops they don't treat us like everyone else. They give us food on broken plates and tea in cracked cups. Sweepers are treated worse than animals," he says. And it's true. They're not allowed to wash dishes in Heera Mandi in case they pollute them, and even the cheapest prostitutes think they're better than the sweepers. Maha rarely sees Tariq because they live in separate social spheres, but when she does, she treats him with a distant and patrician kindness.

Sweeper families are invariably very, very poor. The government gives them contracts to clean a specific area. They call this their "duty"—and it's an impossible task for which they receive a pittance. Dealing with rubbish is considered contaminating and beneath the dignity of most people, so they chuck it out of the windows of their homes or drop it in great piles in the street. It's left there to stink and rot—a feast for the rats—until the sweepers come to collect it.

Sweepers supplement their municipal duties by working on a private basis for individual families, and they do it for the most meager rewards. It's almost impossible for them to change their occupation because of the stigma that is attached to their families. They are shamed by their low origins and their dirty jobs and few people support the rights of poor Christians. They are *kafirs*—unbelievers.

It's four in the morning and still dark, so I'm walking carefully to avoid falling into any of the open drains. The sweepers have no problem negotiating the *galis*; they have developed such a long and intimate knowledge of the lanes they clean.

Sweeper women are working, ankle-deep in rubbish, alongside men. Lots of women work in the old city but they do so in the *purdah* of their homes. They sew, they cook, they assemble things—often for a fraction of what men would earn for doing similar tasks—but they don't do these things publicly. The only women I see working in Heera Mandi are the beggars, the prostitutes, and now the sweeper women. Some sweep with their families and others labor alone—ragged, wizened women, old before they've reached middle age.

In Tarranum Chowk two men are clearing around the addicts lying unconscious on the roadside. They greet Tariq and tell him about their cart: its wheels have broken and it's impossible to push. They don't know what to do—the municipal authorities have told them they have to repair it themselves. Four people are sweeping near the church. Heera Mandi Chowk is wide and always busy: it's far too much work for one person and the sweeper responsible for the area brings his wife and two children to help him. They stop sweeping to speak to us. The father is out of breath and perspiring heavily. "It's four people's work and one person's payment," he puffs.

We walk around the *galis* on the fringes of Heera Mandi. We're on our way to see where Tariq does his "duty." We pass through an especially narrow alley strewn with debris. "This is the musicians' *gali*," he comments. "They're very dirty people—the worst in Heera Mandi."

Tariq has two winding *galis* to clear. On most days he'll get up at three so he can start work at three-thirty. His duty takes him three or four hours, so he finishes his government job before the day has properly begun.

By the time dawn has broken Tariq's brother and nephew have almost finished sweeping the courtyard. They brush the litter into neat piles and shovel it into the government cart. It is a big yellow metal box on hard little wheels, and they take it, squeaking and rattling, to the dump where rag pickers and bottle-top collectors are waiting for them—filthy, thin, and anxious—to sift through the mess.

Tariq wants to know why I'm interested in his work and what I'm going to write about it. I tell him it's important work. We've a different system in England and people from my country don't know much about places like Pakistan—and especially not places like Heera Mandi. I tell him that I'm writing about life in the *mohalla*.

"About good things as well as bad things?" he asks.

"Yes," I reply. I have many things to write about, but the dignity of Tariq and his family will be among the easiest. Their dignity stems not from their religion—although this is something that gives them comfort in the midst of perpetual poverty—but from the fact that they are the only people I know in Heera Mandi who are not caught in the soul-consuming web of the business.

White Flower

The most attractive man I've ever seen is sitting with White Flower on a sofa in the *khusra* house. He has a chiseled jaw, a full mouth, and beautiful eyes. He's the epitome of masculinity—until he squeals and begins to speak in an extremely high-pitched, feminized voice. She's arguing with White Flower about "deck functions," and they rub each other's thighs whenever they want to stress an important point. The handsome *khusra* says that traditional musicians are best. White Flower doesn't agree. She thinks deck functions are best.

"The customers don't want to pay for expensive shows," she explains with her hand in the gorgeous *khusra*'s groin.

The argument is loud and they wrestle, dropping to the floor, their clothes getting covered in dust and dirt. A young man staggers from the toilet. He can't get round them, and so he falls, collapsing onto the floor with his hands over his head.

White Flower introduces me to the youth by pointing in his direction and saying, "He's my husband." The youth is too ill to speak: he's been taking drugs and drinking whisky.

The youth is only about 16 or 17; White Flower is at least 40 and has such a dominant personality that I can't imagine her adopting the role of wife with this teenager. So I ask White Flower if she ever behaves like a husband.

She laughs. "I'm always a wife," she shrieks as if it's too obvious to state. She undoes her hair clips and tips her head back. "Look at my hair."

I've promised Tasneem a new outfit. She's admired one of mine many times: it's blue cotton with pink embroidery and I bought it in Delhi. Her own clothes are a horrible collection of old polyester *shalwaar kameez* and most have holes or tears. I've scoured the bazaars in Lahore for something similar to mine, but there was nothing suitable and I have no time to keep looking. I give her a little roll of notes and ask her to buy her own and explain that I couldn't find the present I wanted. She pushes the notes into her bra and rushes out of the door in panic, her pupils wide and black with nervous excitement. When she returns, a few minutes later, she's a little calmer and she tiptoes slowly and stiffly around the room, trying not to draw attention to herself. It has exactly the opposite effect and White Flower screeches at her to turn on the music.

The room is filling up with a new batch of *khusras*. White Flower picks herself up off the floor and sits in state on the bed. She's important here because she owns the house and because she's a very good dancer. She was born into a traditional brothel family and she's the guru for all these other *khusras*. This means that she manages their work and much of their lives. There are other gurus above White Flower in Heera Mandi's *khusra* hierarchy, but in this house she wields power. She doesn't live here in this gloomy room so much as hold court.

If White Flower is at the top of the pecking order, Tasneem is at the very bottom. When the phone rings, it's Tasneem's job to answer it. When drinks are needed, it's Tasneem who skips to fetch them. When the guru shouts, Tasneem jumps with fright.

Two *khusras* are sitting on the *charpoy* by the open window painting each other's nails. One is about 30 and the other around 20. The younger one introduces me to her friend. "This is my mother," she says.

"And this is my daughter," the older one adds.

It's common in Heera Mandi to refer to people as relatives even when they're not connected either by blood or marriage. Good friends aren't simply friends — they're something more intimate. I'm not Maha's friend: I'm her sister and I'm her children's aunt. It's even more important for the *khusras* to have relationships like this: they've left their families and abandoned mainstream society and it's among the *khusras* — people like themselves — that they find their new relatives and life's meaningful and lasting bonds.

Someone passes a pair of tweezers to the *khusra* sitting next to me, and she begins to pluck her beard absentmindedly as she talks. Determined *khusras* pluck rather than shave. It makes their skin smoother because regrowth is slower and they don't have permanent, thick black stubble. It's a nonstop process and the tweezers are passed from one to another in a never-ending circle.

I'm leaving and Tasneem comes down the stairs with me. She kisses me and pats the rupee notes in her bra and, although her guru is three storeys away, she whispers close to my ear. "Don't tell White Flower you gave me this. She takes everything I have and she'll take this as well."

The Rains

An oppressive humidity builds over Lahore during the monsoon, the dark clouds gathering lower until, every week or so, the weather breaks in a sudden torrential storm, clearing the air for a few fresh hours before the heavy, muggy heat returns again to slow the day. A violent downpour immobilized the city this afternoon. It was preceded by a strong wind that tossed the plastic chairs in the rooftop restaurant into the air, tumbling them over and over and crashing them against the walls of the terrace. Gusts blew through the house, banging the heavy wooden doors, rattling the shutters, and setting dry powdered earth and dust swirling through the streets in pale brown clouds. The drier surface layers of rubbish were stripped from the soggy mass underneath and scattered, sticking to buildings and passers-by. Plastic bags whipped back and forth, some getting lodged high up among the lattice of electricity and telephone wires.

Raindrops the size of giant marbles began to drop onto the courtyard with a hollow slap. They came slowly at first, just one or two, and then a deluge followed. It was warm rain that drenched anyone still out on the streets within seconds. Small boys undressed and frolicked in the courtyard in their shorts. They splashed in huge pools, ran through tumbling streams, and then, when they were tired, they rested on an old, sodden, abandoned sofa in the street. Two young men stood side-by-side in the middle of the courtyard, their faces turned to the sky, their eyes closed, and their soaked *kameez* clinging to their skin. The women and girls watched them from behind their bamboo blinds, keeping dry, as Heera Mandi was cleansed by the monsoon.

Ghazal

Maha continues to rehearse her repertoire in preparation for her trip to the Gulf, and a group of *mirasi*—the traditional musicians of Heera Mandi—have arrived to put her through her paces. One man is playing the *tabla,* another the *dholak,* and a third has a harmonium. Maha brushes her hair, arranges her suit, and applies some crimson lipstick. Her performance must be not only technically perfect but pretty to look at as well.

Maha and her musicians practice for over two hours, producing some wonderful, inspired *ghazals* in this small gray room. In the heyday of courtesan culture the *kothas* of accomplished *tawaifs* reverberated with *ghazals,* a music drawing on Indian and Arabic tradition to create a complex and refined cultural form that blended rhythm with exquisite poetry. *Ghazals* were a type of light classical music enjoyed by elites sufficiently educated to understand the sophisticated verse—the short stanzas containing a universe of meaning in two lines and the beautiful, meaning-laden refrain delivered by a pure and expressive voice. The importance of *ghazals* declined along with the vanishing world of the courtesans, but the genre enjoyed a resurgence when *ghazals* were popularized in Bollywood films. The songs that Maha sings in her cramped rooms are a modern hybrid *ghazal*, halfway between a classical *ghazal* and a *geet*, a popular, more folksy song, but she retains elements of the true *ghazal*. Her voice soars, powerfully but gracefully modulated, and saturated with feeling. She is singing a tale of love, repeating a refrain about unrequited, endless longing. She meets her lover by accident and her heart is forever his, yearning and always destined to be alone. They are heartbreaking but uplifting tales that mix the passions of earthly pleasure with a poetic romanticism. Maha executes them so well and with so much artistic truth because they have a personal meaning to her and to all the women of this *mohalla*. Love will escape the vast majority of Heera Mandi's women, and the only men they will ever meet will be their clients, who may love them for a while but who will ultimately despise them as *kanjri*. In Heera Mandi, *ghazals* are laments and a dazzling, cathartic, public celebration of private tragedy.

Maha and her musicians are happy: they know they're good. The children are crammed in with us. Mutazar keeps draping himself across his mother's legs. Sofiya climbs up the bedding arranged in a haphazard column in the corner and she wobbles uncontrollably six feet above the concrete floor. There's been a fight over an old atlas bought from one of the salesmen in the courtyard and—horrors—the map of Pakistan has been ripped in two. Maha continues singing, interspersing the lyrics with shouts of "Stop it," "Die, idiot," and "Get out." After each ear-splitting, enraged shout she returns immediately to the *ghazal* with complete composure and an utterly radiant smile. The ensemble stops now and again to repeat a section, to correct some point of musicianship or to perfect Maha's delivery. Ama-Jee wades through the bodies in the room, distributing tea and *namkeen* and mopping up spilled drinks and Sofiya's wee. If Maha can perform in these conditions and with such an unruly audience, she'll be a star on the stage in Bahrain.

Ariba, Servant, Slave

There's a strange atmosphere in Maha's house tonight. It's a kind of party, but nobody is jolly. Maha introduces me to a new couple.

"This is Saheen and this is her husband. Saheen is my sister . . . and she's also Ariba's mother."

I look quizzically at Maha.

"Her daughter died a few months ago," Maha says. "And now Ariba is going to be Saheen's new daughter. They're going to Sheikpura tomorrow."

I look at Ariba and she scrambles out of the room. Nisha and Nena give me a warning look, and Ama-Jee shakes her head and lowers her eyes. I follow Ariba. She's not in the bedroom or in the kitchen, but the main door is still bolted from the inside. I find her hiding in the kitchen cupboard. She's rigid: too traumatized to fight and too scared to run, but her eyes are panic-stricken and unforgettably petrified.

I ask to speak to her mother in the semiprivacy of the bedroom and Maha begins her defense. "Adnan wants her to go. He won't pay for her food. These people will give her things I can't. They will send her to school and give her good food and a chance to be away from Heera Mandi. She'll have a better life. She'll study the Quran."

"She won't study the Quran. She'll be a slave," I plead. "She's your child. You can't abandon her." Pakistan has a long and thriving tradition of slavery, and I fear that Ariba will become another addition to an invisible scandal.

Maha shifts uncomfortably. "Ariba is bad. Everyone says she's bad. And I'm going away for three months. Ama-Jee can't watch her all the time."

Ariba's eyes appear around the door frame and disappear again. I know that she trades sexual favors to boys in the streets for a few rupees. I've heard the rumors and I saw her once pressed up against a wall by a boy of 13 or 14 in a narrow, rubbish-strewn *gali* off Heera Mandi Chowk. These boys are getting something from her. And it's a reciprocal arrangement because Ariba receives something from them in addition to the rupees in her hand: she's wanted—in whatever sense that might be.

"Poor Ariba," I say.

"No," Maha interrupts. "Poor Maha! Poor Maha! I have no money. No husband. What can I do?"

"Poor Ariba," I insist. "If you send her away she'll be so unhappy she'll become even more badly behaved. She needs you and she needs her sisters. She needs a mother to love her."

Maha looks at me and the tears run over her cheeks. Her daughter is hovering by the door and hides in the kitchen as I leave.

She will not go to Sheikpura. Two days later I see Ariba waving from the balcony and blowing me a kiss.

Magic

A spiritual session is in progress when I arrive at Maha's house. A *pir*—a holy man—is sitting on the mattress looking serious. He's middle-aged and wears a white *shalwaar kameez*. His hair and beard are dyed red. He holds a string of prayer beads, rubbing them and passing them through his fingers, each one circled by a fat silver ring. Maha sits at his feet, her face scrubbed of makeup and her eyes puffy with crying.

The holy man holds his audience enthralled—except for Ama-Jee, who is standing in the other room, out of his sight, pulling grotesque faces.

There are long silences while the *pir* clears his throat and looks at the ceiling.

"I have seen the sun rise and I have seen it set and I know why." The beads pass through his fingers a little more slowly. "The moon and the heavens are God's creation." Maha cries silently, rocking back and forth in misery. The mystic continues for half an hour in the same vein and then grows silent.

I go into the other room. "What's happening?" I ask Ama-Jee.

"He's waiting for his dinner."

The meal arrives after another fifteen minutes. It is a feast: chicken, *dal,* curried vegetables, and a tower of *roti.* The *pir* polishes his plate and then speaks to Maha in private. He leaves a few moments later and promises to come back tomorrow.

Maha sinks on to the mattress and sobs.

"I knew it. It's black magic. That bitch Mumtaz is using black magic. And my mother's new husband—it's him too. Both of them. Two spells." She takes long, deep, jagged breaths and trembles. "Mumtaz has paid a man in Peshawar to use black magic to make Adnan love her. And my stepfather wants me to die so that he and my mother can take the girls and put them in the business and make a lot of money. He put some magic powder in my *shalwaar* while I was taking a bath." The inexplicable has been explained. And now all Maha has to do is to reverse the black magic and life will return to its happy and ordered state. The cost of repairing her life will be expensive. The mystic will start working on disabling the spells tomorrow and his fee will be five thousand rupees—about $84.

magic, sorcery, and the evil eye help shape the lives of the people of Heera Mandi and Maha's world is populated by jealous enemies, saints, and *jadugar,* or magicians. *Pirs* like the one advising Maha are regular visitors in the *mohalla.* They claim to be the descendants of great Sufis—Islamic mystics—from whom they have inherited special powers. Some are deeply religious; some are sophisticated psychologists; and some are total charlatans. Throughout Pakistan people turn to *pirs* during times of personal crisis, and in Heera Mandi they're in constant demand.

Pirs are thought to use *nuri ilm*—luminous knowledge—a good and positive force drawn from God. *Kala ilm* is the precise opposite. It draws on the negative powers of Satan and demonic spirits and is contrary to Islamic teaching. *Jadu,* magic, sometimes uses the good knowledge. But in Heera Mandi, *jadu* is associated with evil and is linked in popular imagination with Hindus and Christians. Its spells and magical potions have a devastating impact upon minds and bodies because they're really believed to work. The *pirs* who tour the district specialize in unraveling spells and exorcising evil. Spells cast by Hindu magicians are considered the worst; that is the type afflicting Maha and why she has to pay her *pir* so much money.

Possessing better luck and a happier life than your friends, family, and neighbors is likely to invite the malevolent gaze and the workings of sorcerers. Jealousy and envy trigger the casting of spells. When a woman has a new patron, new clothes, and new jewelry, she says her enemies are "burning." They cast an envious eye upon her and those who burn the most fiercely may employ a *jadugar* to do real damage. Magic and spells are used by those outside Heera Mandi, but they have a special potency here: it's the way many cope with living a stigmatized life in a marginalized community.

Maha fears her mother's envy, and she has good reason to worry. Maha disobeyed Kanjar codes: she left Heera Mandi with her *sayeed* husband and her family lost the earnings that were supposed to sustain them. When Maha went to live in a nice house, her family remained in the *mohalla.* Maha thinks, probably correctly, that her mother has never stopped burning at the memory, or at the sight of Maha's daughters who could now be providing their grandmother with a comfortable old age.

Maha also has to cope with the witchcraft used by Adnan's wife. According to the *pir,* and fast-spreading rumors among the local women, Mumtaz pays a Hindu sorcerer in Peshawar to cast spells that keep Adnan a happy and faithful husband. He's given Mumtaz a recipe for a special love potion. Maha is sure it works. Mumtaz mixes her menstrual blood in Adnan's food and drinks: he's bewitched and enslaved to his wife's charms.

The evil eye is everywhere in Maha's world. *Bhuts*—the spirits of the dead—inhabit the earth, swooping through houses and dark places spreading terror and the chill of bad luck. Maha knows where these cold, misty spirits dwell so we rush, shuddering, past the sites of murders and ancient but never-forgotten crimes. And when she saw me fall victim to the world of magic, Maha knew how to fight the spells and chase the evil away. She is always here to look after me.

Down by Roshnai Gate a couple of men sell special *paparhs*—a kind of poppadam that contains roasted coriander seeds and plenty of marijuana. *Paparhs* are wrapped in newspaper and surreptitiously heated over tiny charcoal fires. They're given, as a matter of course, to young girls in cheap brothels who service lots of clients. *Paparhs* can be unexpectedly potent because the Taliban in Afghanistan has released large amounts of opium and heroin on to the Pakistani market and some of this has apparently found its way into *paparhs* like those made by Roshani Gate. I ate some of these *paparhs* in a nearby brothel and washed them down with tea, unaware that they contained more than the usual mild narcotic. Within half an hour, I was living in multiple universes: having conversations with my children, visiting Brazil, buying tomatoes in a supermarket, and driving into a brightly lit tollbooth in France. I stayed in these vivid universes—and in Maha's close care—for days. In the moments when I returned briefly to the realities of Heera Mandi, Maha gave me water and cried and prayed to Allah to save her sister. I woke once to find her circling my body with a bag of meat. Another time it was an egg. Later, I woke to a black hen's head being waved three inches above my face, its beak open, its eyes staring, and its feathers shaking. Each time she took these things to the balcony and threw them with all her might across the courtyard.

When I'd recovered sufficiently to ask her what she was doing, she took me to the balcony and pointed at the house that belongs to Mushtaq, the big pimp.

"He saw you," she said. "That time when you were sitting here and Nisha was checking you for lice. You weren't wearing your *du-patta* and he saw your golden hair and he wanted you. He put a spell on you." She believed that I had overdosed on drugs but was convinced that an evil force had led me to eat the intoxicating *paparhs*.

The meat and the chicken's head that she'd hurled at the pimp's house were meant to undo the spell. I couldn't fathom it. Why, if Mushtaq had desired me, would he want to make me ill?

"He's stupid, *badtamiz*," Maha shrugged. "The spell must have gone wrong."

"Big Love: Big Money"

(COLD SEASON: NOVEMBER 2000 – JANUARY 2001)

I'm so pleased to be back after an absence of three months. The weather in Lahore in November is as perfect as it can be: the days are warm and sunny and the nights are cool. Physical activity no longer induces heavy perspiration and, judging by the number of people on the streets, a large proportion of the Lahori population appears to enjoy walking around the city as much as I do. Today, every rickshaw and *tanga* in town was pressed into the old city and the roads near the cloth market became gridlocked with bodies and vehicles. A van was jammed against the stalls and a rickshaw fought to pass in the opposite direction. Behind them other vehicles became compacted and wedged into the *gali* so that no one could move.

Lahori pedestrians possess a staggering disregard for the dangers of traffic and a

firm belief in their own immortality. They stand in the midst of speeding cars, stroll in front of lorries, dodge rickshaws on the main
roads, and squeeze themselves through minuscule gaps. This afternoon, a round, middle-aged woman complained about the delay and
then spotted an escape route. She edged between a van and a stall
where the shopkeeper was frying samosas in a big vat of bubbling
oil. The women sucked in her stomach to avoid the flames licking
around the bottom of the vat and plopped out of the other side with
her suit spattered with oil and the end of her *dupatta* on fire. She
shouted and beat out the flames on the side of a rickshaw and continued up the lane trailing smoke and the smell of burnt polyester.

A donkey cart became stuck next to a vegetable stall and the animal took the opportunity to stuff his head into a luxuriant pile of
spinach. Donkey carts seem to be characterized by an inverse relationship between the size of the donkey and the load on the cart:
donkeys that are skin and bones and barely three feet tall pull giant
loads and fat drivers. When the jam eased, the donkey strained and
jerked along the road, his hooves making an unsure, sliding click on
the concrete. A mountain of boxes was assembled on the cart, and
the driver wielded an enormous stick over his back. The little animal struggled; his harness didn't fit properly, chafing through his
skin so that he had a long, weeping wound an inch deep in his flesh.

I've never been able to explore the alleys of the Walled City and
find my way back to Heera Mandi without assistance. Countless lanes
and indistinguishable passageways wind between tall and equally indistinguishable buildings so that I soon become disoriented. I realized
I was lost today when I passed the same butcher's shop three times in
the space of an hour. The butcher had arranged goat heads very neatly
according to the color of their few remaining tufts of hair; starting at
white and progressing through shades of brown to black. A dozen
stomachs, like latticed footballs, bobbed in a tub of water beside a
beautifully symmetrical stack of goats' feet. *Paow*—stewed goats'
feet—is a specialty in the Walled City, and people come from all over
Lahore to sample the delicacy in local restaurants. I'm told that it's delicious and have been offered the dish on many occasions, but I cannot

bring myself to eat it. My hosts underline how narrow-minded I'm being by noisily sucking at the marrow.

Vegetarianism is not lauded in Pakistan. Poor people are reluctant vegetarians, and if you can afford it, you eat meat—lots of it, preferably with a hot, oily gravy. Even the *dal*—that staple of the vegetarian diet—is fortified by simmering the lentils with bones and bits of animal. Occasionally I lie and claim to be vegetarian in order to avoid eating heads or sweetbreads, but this fib is always greeted with some surprise, as if I might be deficient in some way—and not just in nutritional terms.

Maha has cooked a special meat-packed meal to celebrate my return. It's perfectly seasoned, the meat is tender, and she adds generous ladles of *ghee:* like Maha, who is wearing her full regalia, lots of makeup and an entire treasure chest of paste jewelry, the meal is an assault on the senses. It's delicious.

I feel as if I've never been away from Heera Mandi. Things alter during my absences. The seasons change, the children grow bigger, the Shia shrine in the courtyard is added to and improved, there's a turnover of girls in the thin pimp's house, and Iqbal is working on a new painting of one of the dancing girls. There's news of marriages, births and deaths, rapes and murders, and fortunes won and patrons lost. But despite all this, the *mohalla* remains unchanged: life presents the same problems and the same solutions to the same kind of women in the same kind of way. Maha might have a new suit and have put on ten pounds in the time I've been in England, but she's still the same soul: veering between rage and joy, cruelty and gentle compassion.

I live in two very different worlds these days and don't feel fully part of either. When I'm at home, I live in a suburb of Birmingham. My children go to school, I shop in the town, I sit in an office at the university, I teach my students—yet part of me remains here in the *mohalla.* In my mind I'm still sitting on the rooftop, walking through the *galis*, and visiting the bazaar at night. I'm watching the life of the courtyard and lying with Maha and her family on the old mattress in her best room.

Yet when I'm here, I also remain an outsider. As a Western woman

I have some safeguard from sexual assaults if I'm veiled, respectably dressed, and maintain an aura of quiet confidence. I'm assumed to be in the protection of a powerful man, one who could wreak vengeance on anyone who should harm me. A high-status Pakistani woman would have the same kind of security. The women of the *mohalla* help me too. They tell customers and the local men that I'm an honorable woman and that no one should go near me. The length of my visits, however, is gradually eroding my status as an outsider and the pimps who kept their distance are moving closer. I'm seen too often, I stay too long with the *kanjri,* and my smart *dupattas* and *shalwaar kameez* have been tattered by the savage Lahori laundry system.

A *dhobi walla* collects the dirty clothes from households, wrapping them in a giant *chador* and taking them to his laundry, where he and his assistants swirl them around in vats of soapy water, squeezing and rubbing until the clothes are clean but noticeably thinned and often tinged gray with dye running from stray black garments. My pretty pastel *shalwaar kameez* return slightly more dingy each time, and so crisp with starch that they crack when bent. I'll be in trouble if I look a shambles: I will look vulnerable. Once the gulf created by my foreignness is bridged, I will be considered fair game and, if I'm on the streets, subject to the kind of dangers—the rapes and beatings—that the women of Heera Mandi are exposed to constantly. Mushtaq, the big handsome pimp, has started to call me over to have a drink with him, and one of Maha's male cousins who works as an agent has been spreading rumors that I drink alcohol and have "relations" with men. Maha has scotched the rumors and taken her cousin to task. Her sister, she says, is a *sharif* woman, respectable.

I'm a little scared when I walk around the *mohalla.* Initially, I'm full of enthusiasm and bravado, but my bravery is soon sapped by this place. I'm always looking for signs of trouble, trying simultaneously to be observant while taking care to avoid eye contact with men. After a week or so I start to spend too much time in my room sending text messages to my family rather than getting out into the thick of Heera Mandi where I'm supposed to be engaging in rigorous academic research. The *dupatta,* which I used to think of as a symbol of oppression, is now my fond friend. I like it and I would no more go outside without it wrapped around my head than I would walk down a street at home in a bikini. In expensive parts of town, where it's not normal for women to be heavily veiled, I feel profoundly uncomfortable when I remove it: for a short while it feels as if something is missing. If this is the way I feel after only a few months, no wonder women who have worn *burqas* all their lives don't wish to abandon them when they are given the freedom to do so.

Still, in the most basic and important sense I'll always be an outsider in Heera Mandi. Like a tourist I will pack my bags and return home. I enjoy a freedom the women here will never know. Maha grew melancholy when I prepared to leave Lahore during the last monsoon. "Louise," she said sadly, "you are like a beautiful bird. You fly here and you sing and make me happy. And then you fly away again. But when you go to another place where you can carry on singing, I'll still be here."

Lal Shahbaz Qalandar

A dozen buses are parked in the Hazoori Gardens, their roofs piled with bedding and carpets wrapped into giant, bulging rolls. They're going to leave early tomorrow morning for the *urs* celebrations of Lal Shahbaz Qalandar, an important thirteenth-century Sufi saint whose tomb lies five hundred miles south of Lahore in Sewan Sharif, in Sindh's lower Indus Valley. The women of Heera Mandi think of him as their special protector. Lal Shahbaz was a *qalandar*, one of the wandering mystics on the periphery of the Islamic Sufi tradition. He is called *Lal*, meaning "red"—because he was said to have worn a red cloak. Like other *qalandars* he's associated with unorthodox behaviors: he took marijuana to deepen his spirituality and connection with the divine, and he was a dervish, dancing ecstatically to reach ever closer to God.

Lal Shahbaz halted his wanderings in an important site of Hindu pilgrimage: Sewan had a Shiva sanctuary where a Shiva Linga (a symbolic penis set within a representation of female genitals) was venerated. This Hindu tradition was incorporated within Sufism and was never completely obliterated. Richard Burton, a forerunner of today's travel writers, visited Sewan in the 1840s and wrote that a girl was dedicated to Lal Shahbaz Qalandar's shrine every year. The parallel with the prostitution of girls in the Hindu temple tradition is unmistakable.

Urs commemorations sound as if they should be sad affairs, but death is interpreted as a union of the Sufi with God: a kind of wedding. For Lal Shahbaz this unity is unusually powerful and poignant; the moment of his death is seen as the moment he met the wife he never had on earth. The Hindu legacy has fed into, and blended with, the Sufi tradition and Lal Shahbaz's personal history to produce an *urs* that has, on its fringes, become associated with erotic license. Women from Heera Mandi go to Sewan Sharif because they think that the *qalandar* will intercede with God on their behalf. They also go to pick up some good business.

The poorest women cannot afford to go to the *mela,* and the women under tight control of their pimps don't have the freedom to do so. For everyone else, the *urs* is a special holiday and one that is anticipated for months. Maha and I are going because we both need to do some serious praying to Lal Shahbaz. I've been told to pray for a husband; Maha wants her old one back.

The more luxurious buses parked in the Hazoori Gardens have darkened windows and multicolored lettering painted on the sides. The older, cheaper ones have seats that refuse to stay upright and rattling windows that either never close or never open. Pilgrims on a tight budget opt for the bus, but because it takes three days to get there, we have decided to take the train instead. It'll be quicker and it's also going to be fun: a twenty-four-hour party. Maha has promised me songs, festive food, and a hallucinogenic drink called *booti.* The train has another advantage over the bus: it has a toilet, and that may well be essential after trying the *booti.*

Maha goes to Sewan Sharif every year or two. She's busy preparing for the trip and she no longer has Ama-Jee to help her. Last month Ama-Jee's long-lost husband arrived in Heera Mandi to claim his wife, raping her so brutally that she was unable to walk. Telling the tale brings tears to Maha's eyes. A few days later, Ama-Jee left in the middle of the night without saying goodbye and taking one of Maha's *dupattas* with her. She had only one *dupatta* herself: an ancient, frayed thing. I remember because she asked if she could have one of my old ones, and then I forgot all about it and never gave it to her. I'm sorry Ama-Jee has gone, but not as sorry as Maha: her plans to sing in the Gulf have been shelved now that there's no one to look after the children.

Mutazar is hampering spiritual reflection and our enjoyment of Maha's stories about the great *qalandar* because he keeps scattering a never-ending supply of *patake* on the floor. *Patake* are essential items for every boy in Heera Mandi: firecrackers that come in packs and range in size and explosive power from the smallest, which are enough to give you a horrible fright, to potent things similar to small bombs that can deafen and injure. Mutazar has the medium variety and is sporting a nasty, watery burn on his nose; a big hole in Nisha's dress has tell-tale scorched edges.

Smoke from the *patake* drifts around the room and Maha is shouting. Mutazar hides his box of firecrackers in his pocket and grins. "I've got lots for the train," he says while the other children gather around to watch him lighting a selection of fireworks. A hissing flash of yellow and green shoots across the room through the open shutters and bursts into a cascade of light in the courtyard. The children cheer and Maha carries on packing. Perhaps it would have been wiser to take the bus.

On the Shahbaz Express

We have enough luggage to leave Heera Mandi forever. Adnan has been persuaded to come with us to Sewan Sharif, and he's arranged a caravan of rickshaws to take our things to the station. We have a carpet, a mattress, pillows and cushions, several rolls of bedding, four bags of clothes and three enormous aluminium pots wrapped in cloths and containing curried vegetables, chicken, and fried meat patties. Maha has been cooking all night, determined that no one should go hungry on the journey.

We're catching the Shahbaz Express, a special train taking vacationers from Lahore to the *mela*. There are about five or six of these trains and each one is packed. We fight our way on board, commandeer two benches, and sort out the luggage with difficulty. Adnan has decided to take refuge on one of the luggage racks above the seats and is sitting among the bedding and bags of clothes, puffing away on something that looks like very sticky liquorice: it's a type of heroin.

Our part of the carriage has seats for about twenty people, but there are over thirty-five of us—mostly young men in Western clothes. They're passing around giant joints. Maha says that many are *mirasi*—musicians—from Heera Mandi. She must be right because they erupt into vigorous devotional songs every few minutes. The songs and prayers become progressively quieter and decrease in regularity as the marijuana and hashish permeate their systems.

Thick smoke fills the carriage—we are all taking drugs this morning whether we want to or not. An old man sitting in the corner starts preparing *booti*. It looks like old grass cuttings. He mixes it into a sludge with water and some youths help him to sieve it. Using a big cloth, they squeeze the liquid into a bowl, dilute it with a bit more water, and pass around the bright green liquid in glasses. It smells like aniseed. Half an hour later they are all asleep, collapsed and lolling on the seats, on the floor, and on each other. They can't have had any *booti* in the next carriage: the women are still singing four hours into the journey and the *tablas* have never stopped.

Mutazar's *patake* would wake the dead never mind the insensible *booti* drinkers. Nisha, Nena, and Ariba scream when he throws them about the carriage. The men think it's funny. They encourage him, giving him matches and yet more *patake*.

Whenever we pull into stations he joins the other boys and youths by throwing them onto the platform to explode among the travelers' legs. Two very big Sikh men became extremely angry at the last station and it was good that the train pulled out when it did. When we slow down as we pass through villages Mutazar and his new friends toss *patake* at the villagers. Perhaps the victims of the firecrackers will guess that this is the Shahbaz Express.

A frothing youth stops Mutazar's assault. He has keeled over and is twitching and vomiting all over the floor and people's feet. Someone says he has eaten a strong, special *paan* and that it hasn't agreed with him. The smell of his vomit doesn't agree with me either.

Maha is force-feeding me from the aluminium pots. She's prepared enough food for the entire carriage, but she expects me to eat most of it. It tastes wonderful, but not in such quantity. And whenever we stop at a station she buys more food from the boys selling snacks at the train windows. We've had sweets, *namkeen*, hard-boiled eggs, potato chips, guava, and a variety of biscuits and fried noodles. If this doesn't stop, I'm going to join the vomiting youth soon.

"Louise, are you having a nice time?" she asks.

I am. I'm looking out of the window. I've rarely seen the Punjab countryside before. It stretches as far as I can see: flat and intensively farmed with abundant crops—potatoes and cotton and a bounty of others. Men are busy in the fields—and women, too, working in little groups in their bright clothes and loose veils.

"Are you bored?" she asks.

I say the view is good, but she doesn't believe me and they all crowd to the window to see what's so good about it. They look puzzled and say I'm mad.

"It's just villages," Maha says. But then, for a moment, there's hilarity: they've spotted some dung cakes drying on a wall.

"Look. Goat shit." They laugh, point, and hold their noses. "How dirty."

After this highlight there's nothing more they find of interest, and they go back to talking and fighting among themselves in the little cramped space that's our part of the carriage. It's as if they've moved the closed, bounded world of their house in Heera Mandi to the train. They're quite sure it is several cuts above the rough, country-bumpkin world of the villages we're passing through.

The carriage has no air-conditioning, so we have to keep the windows open to cope with the hashish smoke. Sand and finely powdered soil swirl around us. An elderly woman—the wife of the *booti walla*—has taken refuge from the dust and the eyes of the men by covering herself in a big blanket. The only part of her that is visible is her toes. She's motionless and a silvery layer of dust and dirt is building up on her black blanket so that she gleams in the sun.

Maha is getting irritable. Maybe all this traveling doesn't agree with her. Mutazar and Sofiya keep clambering up the ladders at the side of the seats, wobbling perilously on the racks above us, and scrambling down again, frequently missing their footing on the thin metal rungs. Ariba has done something wrong, although I can't work out what it is. I think her very existence is enough to incite her mother's wrath. Maha takes a shoe and starts beating Ariba with it so hard that, after a while, even the semicomatose *booti walla* grows uncomfortable and shouts at her to stop. Ariba weeps for an hour, Maha simmers with anger, the other girls are silent, and Adnan is unconscious.

night has fallen and the light has been switched off in the carriage. We're stopped at another station and Maha is buying tea from a boy on the platform. He passes the cups in through the window. They're made of paper and have been acquired from Pakistan International Airlines. Sofiya lunges for one but it's too hot for her and she drops it, spilling boiling tea over my sandaled feet. It's acutely painful and I'm worried that my topmost layer of skin has been scalded away.

Maha is in a panic about the accident and when we pull out of the station she rummages around in the luggage. She smears something thick and white over my feet. I can't see what it is in the dark but it does feel remarkably cooling. Then the smell reaches me. It's minty.

"Maha," I say, "it's toothpaste."

"Yes," she replies confidently. "It's special Pakistani medicine. It's Colgate."

S pending twenty-four hours on a hard bench enveloped in hashish fumes is testing my endurance. The young men say that we're nearly there. They have been saying the same thing for two hours and I no longer believe them.

We're traveling through the arid scrub of Sindh: squat buildings and the occasional splash of green vegetation dot the dusty fields; a few barren hills rim the horizon. And then someone says that he can see it—he can see Sewan Sharif. Everyone scrambles to catch a glimpse of a town rising in the midst of the desert. At its heart is a giant golden dome. The carriage rings with shouts and ecstatic songs. The young men are crying and praying. Maha has her hands open in supplication, and tears are making little trails through the dirt on her cheeks.

Sewan Sharif

The floor of the train is covered in debris: food, paper, plastic bags, and burnt *patake.* Adnan is ankle-deep in the rubbish and sleepily directing porters and some of the young men to move our luggage. The station is on the edge of town next to sand dunes where hundreds of poorer pilgrims have set up camp in tatty tents and awnings. We hire a couple of carts pulled by tiny, skeletal donkeys to carry our luggage and we ride into town in a *tanga.*

Enormous tents line the main roads, full of stalls selling kebabs, traditional sweets, and souvenirs: pictures of Shahbaz Qalandar and models of his tomb. The place is packed and the roads are congested with people and traffic. Most of the transport is horse- or donkey-powered and there are very few cars. I can't see any foreigners—just local men. Maha squeezes my hand. "So many men," she chuckles making an elaborate performance of arranging her *chador*. "They're all looking at us."

Sewan Sharif looks biblical in the manner of the pictures I remember from my book of *Children's Illustrated Bible Stories:* low buildings with flat roofs sit in the sand. The houses are built around a courtyard and the windows all face inwards. Each house is like a little fort and no one can see inside.

Only a few of the roads are paved and most of the sewage and drainage system is aboveground. Channels full of rubbish, shit, and wastewater run alongside the buildings. The butchers are doing a brisk trade selling meat for the pilgrims' celebrations, and the gutters still run red a hundred meters from their shops. Two men weave between our *tanga* and the crowds, their wheelbarrows full of freshly slaughtered goats skins—so fresh that they move like liquid in the barrows.

The *tanga* can't cross the center of town, so we have to walk through the press of bodies. The *booti walla* and his wife—minus her shroud—are with us. I don't think Adnan knows where he's going. We've been walking for ages and peeping into houses. Perhaps he's trying to work out where we're supposed to be staying.

Lots of Sewan Sharif's residents rent out rooms to pilgrims during the *urs* commemorations. They divide their courtyards into sections and erect tents. The cheaper ones really cram the guests in: there must be a couple of hundred in some, and the makeshift toilets that are made by hanging a few sheets around a hole in the ground are already beginning to stink.

A substantial and inviting house stands at the top of a steep path leading down to the central bazaar. The courtyard is picturesque: shaded with trees and vibrant with pretty flowers. The walls are recently whitewashed, and parts of the floor are set with blue-and-white tiles painted in intricate, geometrically perfect Islamic designs. Colorful awnings and tents divide the courtyard into charming holiday homes. We sag in, covered in dirt, and the *booti walla's* wife sinks onto a wall. A fat man—possibly the owner—doesn't like the look of us and we are shunted out the door. He wants a better class of pilgrim in his house.

Our lodgings are nowhere near as luxurious, but anything is welcome after the train and after fighting our way through the melée in the bazaar. The owners of the house are a pleasant Sindhi family who have set up camp in their own courtyard so that we tourists can move in. Everyone staying in the house is from Heera Mandi: us, a few women, and lots of musicians from the *mirasi*—a group that the Kanjar think are far lower than them in the social order. Maha says it's not ideal but that we'll manage. We have a room to ourselves and we've spread the carpet and beaten out the worst of the dust. The mattress is down and the cushions have been spread artistically around the room. Maha sighs with approval and states that she's going to have a perfect holiday.

A door leads into the courtyard and we are told that we can share the bathroom with the ladies of the house and the other female guests and their children. It's a strange bathroom that has been added onto one of the sides of the courtyard: clean but very primitive. You walk up some steps, squat over a hole on a podium four feet in the air and, because there's no ceiling, an assortment of interested viewers can peer down from the higher surrounding building. There's no door either—only a thin curtain that dances in the breeze.

Far more men than women are holidaying in Sewan Sharif. In our lodgings most women are staying in their rooms or are busily engaged in some task, like laundry or cooking. The small kitchen—four feet by five feet in size—is generating tension among the visitors. One little gas stove with a single burner sits on the concrete floor. It takes a long time to cook anything because the gas supply is low and there's always a line of people impatient to cook. All the washing-up is done in the courtyard under a tap, but this is also creating trouble: others, it is claimed, are leaving the place in a mess. Bits of vegetable peelings are lying all over the floor.

Ariba is upset. She's the only one who doesn't have new clothes for the *mela*. Even I have a new outfit: Maha bought it for me. Nisha and Nena return from the bathroom looking pretty. They've scrubbed up well and are dressed in bright new *shalwaar kameez*. Sofiya and Mutazar have been washed too and are now naked as they cavort around the room. Sofiya is told to keep her legs together and not to sprawl all over the place like that. Only *gandi* girls do that sort of thing.

It's getting dark and Maha and I are going to visit Lal Shahbaz's mausoleum. I hope we can find our way back through these streets. They all look the same: full of sand, piles of rubbish, and window-less buildings of crumbling brick.

The bazaar is packed with pilgrims, most of them highly charged young men. Maha is saying a prayer beneath a giant pole sur-rounded by candles when she's interrupted by a sudden surge in the crowd, and we are pushed on toward the mausoleum. Too many people are squeezed together here and I'm worried about the mass crushings that are sadly common at religious events like these. The men are working themselves into a frenzy: assembling in groups, waving flags, jumping up and down, and shouting incoherently. They surge forward from the bazaar, through a passageway, and into the square directly before the mausoleum.

Maha and I stop in the passageway to buy flowers from a stall to throw on the shrine. The flowers are beautiful: pink, heavily scented, and strung on a thread like a necklace. We lean against the wall of the passage as a wave of men push through. Moments later we are not leaning against the wall but pressed tight against it. Some of the men's faces are angry; some are frightened and shocked. They push this way and that, scaring others who push too. I think I can escape by climbing over the wall, but I can't lift my arms because the press is so intense. My bangles snap against the bricks and the glass cuts into my arms. The pressure is so intense that I can't breathe and I think my ribs will snap as well. Maha's eyes are full of panic as we are propelled slowly toward the mausoleum, scraping against the wall until we are ejected from the passage and can breathe once more.

Hardly an inch of space is left in the courtyard that's not thick with pilgrims. We've removed our shoes to show respect: we left them with an old woman and I wonder if we'll ever see them again. Many of the pilgrims are in a trance. On the right-hand side is the ladies' area. Some have babies and young children. I recognize a few of the faces from Heera Mandi. Many are praying. Dozens of the female devotees are honoring Shahbaz Qalandar by imitating the illustrations of him that are sold in the bazaar: they sit cross-legged or twirl around, undoing their hair and tangling it so that it swings from side to side in matted strands.

Most of the people in the courtyard are men: poor peasants and laborers. Their fervor increases the closer they get to the mausoleum: dancing with their arms in the air and shouting so loudly that I have to scream to make Maha hear me. A few of the men are spinning like dervishes, and Maha gestures toward them. "They're like Shahbaz Qalandar. It's the hashish," she says.

The men gather in tight groups to chant and build up a head of steam before pouring into the mausoleum. Nothing stops them: not me, not Maha, not the thin old man with a stick who stumbles over the doorway and is trampled by the charging youths. Someone grabs the old man and pulls him along the floor through the racing legs. He's too scared, or too old and rigid, to curl his body to protect himself from the blows. I catch a glimpse of his shocked face among the feet and pray he will emerge alive from the stampede.

The mausoleum is circular, the dome rising high above us. Wealthy men—portly landlords from Sindh—and their henchmen watch from a balcony running around the edge of the mausoleum. The tomb in the center of the frenzy is covered in silky cloths, flowers, and a golden turban topped with a plume of feathers. Around the coffin a silver-plated wooden frame is strung with tinsel. Four men stand inside the frame alternately handing out flowers or sweets and beating back the worshippers with *lathis*, long bamboo canes. Scaffolding has been erected around the wooden frame so that the pilgrims don't overrun the tomb. More guards swing on the scaffolding, fending off the crowds with *lathis*.

This act of worship is a frightening outpouring of desperation. Some of the men look as if they're in ecstasy, but most seem in excruciating pain. Those who can get near the frame touch the silver. They kiss the tomb, beseeching it and praying to it. Old men can't get close enough. They're pushed aside by younger, fitter, bigger men, and angry shouts and scuffles break into vicious fights. The scaffolding creaks and swings to and fro with the weight of so many people. It looks as if it's going to topple over into a bloody mangle of iron bars, limbs, and tinsel.

The women have their own section at the back. The scrum is not as rough here, but the women still grab at the clothes of those near the tomb and pull them down and back into the crowd so that they can seize their place and edge themselves closer to Shahbaz Qalandar. We're in a crush of people ten or twelve deep and Maha is dragging me up toward the tomb. Many of the women holding on to the scaffolding are distraught. A young woman with a disfigured face, incapable of expression, is clinging to the inner frame. She's been burned so badly by fire or by acid that she seems to be wearing a shiny mask. Her eyes are permanently open, and I don't know whether she is crying now because she's happy or sad or whether she always weeps.

One of the men inside the shrine gives me a brightly colored shawl edged with gold thread. It's been draped over the tomb and has absorbed some of the Sufi's special powers. The other women tell me I'm lucky: nice things will happen to me in the near future. The cynic in me smiles — I'm an atheist and an academic and I'm not supposed to place faith in the illogic of the supernatural — but, secretly, I'm really looking forward to all those nice things. I sink back into the crowd and tie my lucky shawl around me for safekeeping.

I might need the blessed shawl sooner than I thought. Hundreds of men chant in the courtyard, and a deafening roar signals their charge through the doors. There's no crowd control and I'm starting to panic. We can't get out. The squeeze becomes tighter. Total confusion descends. I've become separated from Maha, but then, as the tide of bodies ebbs a little, I can see her hanging over the silver-framed tomb. She's waving angrily to me to come back and pray some more. When I'm close enough, she tells me to get busy and ask Shahbaz Qalandar to answer our prayers. "Pray that Adnan will love me," she instructs. She's crying again. "Pray he leaves his wife. Pray she gets cancer and dies."

A group of men are tossing sweets high into the air. They've been donated by people who want to gain religious merit and have been blessed by placing them next to the tomb. They're like small, hard marshmallows. Some stick in my clothes. A few of the old women have caught a supply in their *dupattas*, but most find their way onto the floor, where the pilgrims scramble to scoop up the treasures. The worshippers eat a few and pass the rest on to others, increasing their potency as they are transferred from hand to hand. Dates, nuts, and popcorn make the rounds too. I've a ball of assorted sticky things adhering to my palm. I can't face the thought of putting it into my mouth after the ingredients have been rolling around the floor. I jump in surprise when a young woman tells me in perfect English that I have to eat something: it's good and lucky. I try a bit and console myself with the thought that the coming year will be a memorable one. I have my blessed tinsel-fringed shawl and I have eaten saint-infused sweets. Maha opens my mouth and checks to see if I've swallowed. She's satisfied. "Next year when we come to Sewan Sharif you will have a husband," she declares.

Hot Dogs

The tea shops and restaurants in the bazaar are packed. A group of rich men are relaxing in one of the better establishments. They all look the same in their matching white *shalwaar kameez*, big moustaches and chunky gold watches. They're watching the pilgrims and Maha is watching them.

The festival sparkles in the twilight. You can't see the dust and dirt in the soft glow from the candles, fairy lights, and low-voltage bulbs. The stalls are piled high with mountains of special foods: barbecued meat, fried pastries, and pyramids of *halva* decorated with pieces of wafer-thin silver leaf. We're accompanied on our walk by a big, old, black dog. He must have escaped from somewhere because he is trailing a long chain. He likes me and won't leave me alone. I like him too but Maha thinks he's about to attack us, and she screams with blood-chilling terror whenever he plods behind us snuffling around our feet.

We're looking for a dress for Ariba. I want her to have something new and pretty like the other girls, but we can't find anything in the market. The children's clothes are made of cheap brittle lace, like little Western-style bridesmaids' dresses with puffed sleeves and big satin sashes. The ladies' sizes are a more customary design but far too big; they're made for short, square women. We've been wandering around for ages and despairing at the lack of sophistication. This place makes Lahore's Babar Market look like the center of the international fashion world. We decide on something in pink and burgundy with tasteless false laces on the front. It's the best they have. It'll be too big but at least it's new and at least it's clean.

Back in the room it's party time. Mutazar has borrowed a ball from a boy staying in the next house and I'm playing catch with Nisha and Nena. They're standing against the opposite wall shrieking with enjoyment. They're teenagers, but they've never played catch before and lack any sense of coordination; when they throw the ball to me it flies in any direction. Sometimes it hits the wall behind them. We've been playing for half an hour and they have only caught it twice. They don't want to stop, but we have to put the ball away because it's time to eat. Maha has cooked lamb smothered in chilies and Adnan is bringing a big pile of *rotiya* from the bazaar. Nisha is asked to chop some cucumber for the salad and Maha sorts through a bag of vegetables, pulling out a *mooli,* a long radish the size of a giant carrot.

"Look," she shouts, waving it around, "it's a big white Pakistani penis." She passes it to Nisha to slice but the girl picks it up gingerly with two fingers as if it is carrying a nauseating disease. She bares her teeth, takes hold of the *mooli,* and snaps it firmly in two, flinging the pieces across the room. Her mother stares at her open-mouthed, and we all fall silent.

"What do you eat in England?" Nena asks to distract us from the broken radish. I tell them how old-fashioned English food is revolting: boiled vegetables and meat without spices, but that it's better now — we eat tasty foods from all over the world.

"In some places people eat pigs," Nena states and everyone makes retching noises. Pork is forbidden food in Islam: it's dirty and unholy.

"And people in America eat dogs," Nisha adds.

There are more retching sounds, and I'm puzzled.

"Yes," Nisha insists. "It's true. I've heard it on the television. They eat hot dogs."

The Booti Walla's Sarangi

It's seven in the morning and the room is still. Someone is playing the most haunting, magical music in the courtyard, soaring from high to low notes, from sadness to joy and back again, in a sweeping, melodious stream. Maha whispers that it's the *booti walla* on his *sarangi,* a stringed instrument played with a bow that is used in northern Indian classical music, often in association with kathak. The grumpy old *booti walla* describes episodes from his long life through the sound of his well-loved instrument, the *sarangi* so much a part of him that it has become his voice.

I'm lying next to Nisha. She'll probably sleep for hours. She's close to me, and I can feel her breath upon my face. Her crippled arm and hand are twisted toward the ceiling, framed in the early morning light angling in a few bright shafts from the gaps in the warped shutters. She stirs and mumbles, moving closer to me so that her mouth is pressed against my ear.

"Please tell my mother that I want to sleep." She adjusts her arm so that it lies across me; so light and fragile it feels as if it will snap if I try to move it.

"Louise Auntie," she whispers. "Help me. I don't want to dance in the bazaar." Her eyelashes are brushing against my cheek, and I'm beginning to understand Nisha's fights with her mother and her reluctance to take her medicine. Nisha doesn't want to be healthy.

She wants to be a collection of deformed bones whom no man will buy. This emaciated girl isn't being stupid or disobedient: she fears being a woman. Nisha doesn't want to be desired. She'd rather take her chance on dying.

"Full Value"

The kitchen is beset by hygiene problems and the *booti walla*'s wife has been accused of spending far too long in there. But this morning, for once, it's empty, and Maha and I have seized the opportunity to clean a space and prepare breakfast. We're having egg fried with onions and green chili, and rice cooked with yesterday's leftovers — the *salan* (gravy) and the well-sucked lamb bones. Maha is in a spiritual mood. She's talking about the magical powers of Shahbaz Qalandar and his protection of women. The abilities of this saint are awesome; she says that he flew like a bird all over the world and then zoomed right up to heaven.

Adnan staggers out of our room. There are dark circles under his eyes, and he seems to have grown very old. He shuffles and shakes his way across the courtyard. The men don't have their own bathroom: they use the open drains that run along the road on the outside of the house.

I tell Maha that the hashish and the heroin are playing an important part in their relationship troubles. She doesn't agree.

"Drugs are good for sex. Hashish is good, and cough medicine too. It stops your *kusi* hurting and you can go on for three, four, or five hours."

She stirs the rice very slowly. She's thinking.

"Louise, what shall I do? My marriage is finished. Next year Nisha and Nena will have to go to the bazaar. They'll earn a *lakh* of rupees to go with men. Then our troubles will be over. What else can I do? Tell me?"

I don't know what to say. I haven't any good ideas. The girls have no education and few skills.

"I'll have to wait until Nisha is better," Maha ponders. "She's too thin now."

"How do they feel about it?"

"They know. It's okay. Nisha and Nena are nice girls. But Ariba . . . she's different. She's like me. She has a good body and she's very strong. She'll be good at the business."

A noise just outside the kitchen catches our attention. Nena is sitting against the wall, rubbing soap over last night's dirty plates. She's been listening to us. She looks up at me and then down again at the plates. She doesn't say a word and there's no emotion on her face.

Some big hot chilies are frying in the pan and their heat makes me cough.

"Business is good here," Maha announces. "Lots of rich Sindh men come to Sewan Sharif. They see the singing and dancing and they pay good money. A few years ago I met a rich man from Karachi. He was big in Customs. He paid for me to stay for two days in the Pearl Continental in Karachi. I went on a plane from Lahore. I got fifty thousand rupees [$843] a night."

"So much?" I question.

She grins, "He got full value."

She's teasing me and her eyes are laughing. "Let's go to the bazaar now and find some men. We'll get fifteen or twenty thousand rupees [$253 to $337] at a time easily. Shall we go?"

Perhaps she's joking and perhaps she's not, but at least she's laughing. It makes a nice change.

Depilation

I'm never alone, not even for a second—not even in the bathroom. I'm followed in by several children and an old woman. Two girls from Heera Mandi appear. They're going through my sponge bag. The soap from London is outstandingly popular: it's been pressed against every nose. They take out and inspect tubes of cream and bottles of conditioner and shampoo. They examine the deodorant and the old woman asks, "What's this for?" She wants to see it put to use.

I have a bucket of cold water for my bath and an audience of eight. I send out all except the youngest children, but everyone drifts back in, bringing others as well.

And now here are Maha and Nena. Maha is looking at me as I stand in the open-air bathroom, naked and freezing, with shampoo in my hair.

"Louise, are you all right? You've been a long time."

I want them to go away but instead they stay to analyze and discuss me. They're interested in the color of my skin. I can hear them saying that I'm all white and pink and that I'd be good in the business. I feel like a specimen in a zoo. And there's something else that they are whispering about, something they don't think is so good.

Maha looks a little embarrassed. "Louise, you're not *ɗaf*." She's talking quietly so as not to shame me. "Don't you clean yourself— down there?" she asks, slightly scandalized.

I don't know what she means. She's just watched me have a thorough scrub of every part of my body.

She persists and eventually I understand. To be a clean and civilized woman in Heera Mandi you must have no trace of body hair. Pubic hair is disgusting: it places you on a par with the animals. Every decent *nachne walli* has a tub of hair-removing cream and the sharpest of razors.

"Do English men like women when they're like that?" she asks in disbelief.

Ariba is rarely with us. She has taken to lying on an old *charpoy* in the messy compound where the Sindhi family has erected its awning. She's disappointed: her new dress is far too big. It hangs on her shoulders. You could fit two Aribas inside it. I can't do anything but promise her a nice new outfit once we are back in Lahore. But by then it will be too late to matter.

Today—again—Ariba will be the only one at the party without a new dress.

The Old Chicken

The room is claustrophobic and Maha is in a deep, foul-tempered depression.

"I hate life," she moans. She's peeping out into the courtyard. It's dark and Adnan is sitting by a lamp, smoking the gooey liquorice with his friends. We have split into the segregated male and female worlds.

"Look at him," she seethes. "He's like an old chicken." She gripes on. "Louise, I'm going to kill myself. I'm going to take poison and die."

I tell her not to be silly, but she continues. "I'm going to get some kerosene and some matches and burn myself. There's nothing left for me."

Her children sit in silence. Nena appears to be on the point of tears, Sofiya is gawping, Ariba seems bored, and Nisha looks as if she'd like to light the match herself. They've heard all this before.

A musical group sashays into the courtyard with crashing cymbals and pounding drums. We put on our *dupattas* and watch from the window as they jig about the courtyard collecting ten-rupee (ten-cent) notes from the men. Maha forgets about the suicide plan while we watch two male dwarfs dancing around Adnan and his fellow smokers. Dressed in ankle bracelets and yellow satin *shalwaar kameez*, the tiny men dance like women. Everyone's attention is caught. Even Maha is laughing. But then she sighs. "Agh, poor things. They're like children but they're men. It's so sad. Why does God give some people good luck and some people bad luck?"

She's crying for the little men—and for herself—as they hop out of the courtyard with their drummers and twirl into the next house holding a fistful of rupees.

Lucky Irani Circus

Today we're going to the Lucky Irani Circus. There are food stalls and sideshows and we pay five rupees (eight cents) to enter the *ajayabghar*, which is not so much a museum but a display of tricks done with pictures, mirrors, and models who have their heads or limbs poking through pieces of cardboard. A severed head has been made from papier-mâché and a woman's head is stuck on a snake's body. It's not convincing: the snake is made out of a roll of stuffed brown plastic and it wriggles whenever someone pulls on a cord. There's a constantly running tap and an empty box from which a variety of things are made to appear. The false bottom of the box flaps suspiciously and the man performing the trick fumbles around in the secret compartment. It's so bad it's hilarious, but then I note with discomfort that no one else is laughing. I hope I haven't spoiled the show for the rest of the audience.

Outside the turnstiles half a dozen men with shocking deformities are lying on the ground starring in a genuine freak show while their promoter collects money. One man's hand has swollen to five times its normal size. He rests it on a cushion. It looks like one of those giant foam hands people wave at sports events.

Back at the lodgings I've been left in charge of the children. Adnan and Maha have both gone somewhere, although not together. Two children from the room next door have joined us. I sit on the mattress while a war rages around me. They're screeching and laughing and tumbling over the floor. Mutazar is being vicious and using the cover of a play session to inflict serious damage on anyone in range. Then the door opens and Maha is back. Thank God. The room is transformed. Nena is tidying, Nisha sits in silence, Ariba leaves for her *charpoy* in the courtyard, Sofiya eats popcorn, and the two children from next door go back home. Only Mutazar continues to rampage.

"Mutazar," Maha shrieks. "Don't beat your sister with that knife. Bring it here. I need it."

By evening Maha is hyped up. She's been singing and dancing in the room for hours and I'm tired of being an appreciative audience. At half past one she's still going strong. I feel ill and am trying to fall asleep, but she's lying beside me asking about Adnan. I'm in no mood to talk through this all over again. "Louise," she questions and nudges me with a hard finger, "does he love me?"

I bite my tongue. I want to say that I won't love her either if she keeps on poking me with that cattle prod of a finger. If I were Adnan I wouldn't love her if she kept on with the dramas and the fights and the constant singing. Where has my lively, amusing friend gone?

Taxi

Adnan has stumbled through the door and half-fallen over Ariba. The light is switched off and we've been asleep for an hour or so. He is mumbling and trying to sort out the blankets. He lies next to Maha. She puts her arm around him and I turn to the wall, uncomfortable by being privy to intimacy in such an overcrowded space. Maha whispers softly to Adnan. Whatever she says makes him angry because he stands up again, damning her with the cruelest words. She's a dirty *kanjri* and a cheap *taxi*—an extremely insulting word describing a prostitute who has so many clients that she's like a taxicab, in which countless men will take a ride. Adnan snatches the blanket covering her and staggers off back to the courtyard. In the dark I can see Maha's silhouette shaking as she sobs in silence.

At four in the morning Maha is pulling me to my feet. "Louise, wake up. We are going to the *mazar*." I'm shivering, my head is hurting, and I can barely swallow. I don't want to go to the mausoleum, but she is propelling me along the streets and I don't feel well enough to argue.

At the *mazar* Maha plunges into the crush. She's shouting something to me, but I can't hear properly: her voice seems far away.

The noise of the crowd reverberates around the building and around my head in a fuzzy echo. I've taken refuge by the rear wall of the mausoleum and hope that Maha comes back soon. When she returns from rubbing her face all over the silver on the tomb, she's still no calmer. We walk home as dawn is breaking and I know it's going to be a horrible day.

The Musical Director's Wife

The tension in the room is unbearable. I lie next to Nisha watching insects crawl into the holes in the plaster of the wall. Maha and Adnan are in the middle of a vile row. Maha is accused of looking at another man staying in one of the other rooms. He's a musician, quite young and only moderately attractive, and he has very little money. I can't imagine Maha would be interested in him. She is supposed to have looked at him from our room as he lay in the courtyard smoking with the other men. Adnan thinks it's an unforgivable betrayal: she didn't keep her eyes in *purdah*.

"You are *kharab*"—a spoiled, rotten woman—Adnan declares so loudly that the whole house will be able to hear.

Maha is distraught. She's been crying for hours and pleading, "I wasn't looking at him. I promise. I swear I didn't look at him." Adnan doesn't want to drop the argument. He's threatened *talaq*, divorce. Maha has only stopped weeping to remind me that Adnan doesn't pay her much. "He gives me three hundred rupees [$5] a day—on the days he does come. It's such an insult. I'm a ten-thousand-rupee [$169] woman."

We have a visitor this afternoon: the wife of the director of an influential musical group, who is also staying at the house. She is in her early forties, slim, and well dressed. She takes the part of a lady of refinement with great seriousness, and I smooth out my crumpled clothes and try to lie a little more elegantly on the mattress. She's well-groomed and possessed of sense: she came on the first-class train for a day trip and she doesn't have thick black lines of dirt under her fingernails.

Maha is being charming. "How do you keep so young and slim?" she asks the musical director's wife. "You've got a figure like a girl."

She laughs and gives us tips on dieting: don't eat too much and avoid large quantities of bread and rice.

"Your hair is beautiful," Maha comments. "Isn't it wonderful?" she says to Nisha and Nena, and they both nod and murmur over its luster.

The director's wife tells us about her busy life, her clothes, and her children, and we are like the best of friends sharing gossip and congratulating one another on our little successes. She leaves after drinking some tea. Someone is taking her to Karachi in a car. I want to ask if I can go in the car too because, compared to Sewan Sharif, even Karachi—that big dump of a city—seems like heaven. She wafts out of the room in a cloud of perfume and chiffon *dupatta,* and Maha snorts.

"She's got a face like an old woman. It's like this." She sucks in her cheeks and pulls her chin with her hand to make herself look drawn and pinched. The girls agree and laugh.

"Do you think she's got a nice face?" Maha asks me.

"Yes," I say.

"But she's got no hips." Maha adds, "She's like a boy. She's an old woman. She's finished."

"Bitch-Man"

We've been here for days and the town is beginning to feel quieter; many of the pilgrims must have left. There's no longer a dangerous crush in the mausoleum, and Maha and I go and hang over the tomb and pray for husbands for one last time. The sudden reduction in visitors allows us a good view of the town. It's carpeted in rubbish. A yellow stream of urine runs down the road and into the bazaar, where it pools, green and frothy, between the tea shops and then gradually filters down a congested hole and into the sewers. Four middle-aged men are relaxing on chairs in the street and enjoying the sun. They look like residents of the town and seem happy to have reclaimed some peace, now that the pilgrims who bring them some seasonal income have departed. Less than six feet from them a pile of nine or ten goats' stomachs and intestines are puffing up as their contents ferment in the winter sunshine.

I can't wait to leave, if only because I can visit the doctor in Lahore. It's Saturday afternoon, a week since we arrived, and we're sitting on the steps of the station waiting for the train. The platform is just a bit of concrete and dust next to the track. Adnan spreads the carpet and we sit down on it while the other passengers walk over its edges, trampling it in the dirt so that I can no longer see the pattern. Maha and I amuse ourselves by searching Ariba's hair for lice. There are enough to keep us busy for hours, but Adnan tells us off. It doesn't look respectable to be hunting for lice in a public place, so we wait at the side of the tracks, by the conical piles of shit that have been dropped from the toilets of stationary trains, and we preserve our honor.

Other women from Heera Mandi are at the station too. A family of pretty girls carrying vanity cases and dressed in cheap but well-laundered clothes are escorted by their mother, an aunt, and a man who acts as their helper and chaperone. I don't know how they manage to walk in their big silver platforms without breaking a leg. Another group of women who aren't wearing *dupattas* look even less sophisticated. A few have their hair streaked strawberry blonde. Maha sniffs and whispers, "Five-hundred-rupee [$8] women."

Dusk is falling as the packed train pulls out of the station. I remember many of the faces from our journey to Sewan Sharif: the *booti walla* and his wife, lots of the musicians who shared the house and made such a giant mess, and—disaster—the man whom Adnan believes to be the object of Maha's passion. He's looking relaxed and oblivious to the scandal, wearing a rather fetching blue shirt and smoking hashish. Adnan is angry and Maha chokes with despair about the horrible, dirty "bitch-man." He doesn't look dirty and horrible, but Maha's re-vulsion is designed to pacify Adnan. She doesn't know where to look to avoid being called a *kanjri,* and so she insists that we section off our part of the carriage. We tie sheets and shawls to the luggage racks. They swing with the movement of the train and blow about in the wind from the open windows, but here in the seclusion of our sheets we are safe: no one can see us and we cannot see anyone else.

Fighting with Maha

By morning Maha is spoiling for a fight. I've been looking out of the window at yet more fields.

"Louise, are you boring?" she asks in a bad mix of Urdu and English that I translate equally badly.

"No, I'm not bored," I reply.

"No, you don't understand. You're boring. You're always looking out of the window and you're not talking to me."

It's a great start to a conversation and we sit in angry silence. "Louise, you're a mental case. Where's your gold?" She pauses and points to my arms and neck. "It's really bad. Where is your *izzat*? If you have no money and no gold, people will think you have no *izzat,* no honor." She's concerned, but she's getting at me too.

"Look at your watch. How much did that cost?"

I've had it years. It has a little leather strap.

"It's *kharab,*" she insists. "It should be gold. You need bracelets for your daughters, and then, when you die, they can have your gold."

Maha has never spoken to me like this. "You write books and you have a good job, but you're really stupid. You have a young face. You have a good body. You have white skin and golden hair. You can make a lot of money and you can enjoy yourself. Find a rich man and then say you want jewelry. You need bracelets, earrings, and a necklace. All gold. Remember, big love: big money. And then you want diamonds. Promise me, Louise. Do it. Don't be stupid. Think about your children."

We were on that train for over thirty hours. Thirty hours of hashish and *patake* and food and fights and swinging curtains. Heera Mandi has never looked more lovely. When we open the door to Maha's house, we may be greeted by a truly foul smell from the toilet and a dozen startled rats who have settled in during our absence, but it feels like home. It's four in the morning and the popcorn seller is still doing his rounds in the courtyard. Mushtaq, the big pimp, is sitting in his house surrounded by his friends while he polishes and plays with his revolvers. Adnan has returned to his wife, the children are asleep, and I'm pleased to be back.

Maha isn't so happy. She can't sleep, and at seven in the morning she's filling tubs of water to wash the clothes. She has her arms in suds and she's crying, *"Lun, lun, kusi, kusi"*—dick, dick, cunt, cunt. "That is my life."

Tasneem's New Home

Tasneem the *khusra* no longer lives with White Flower. She says Tasneem is ill and has gone to Peshawar to pray for a miracle cure. I know she's lying because I spot Tasneem rushing to the PCO—the public call office, a shop where you buy phone calls. She falls on me like a long-lost and dearly loved relative and the bazaar comes to a standstill around us.

She's moved to a new, much nicer, and lighter *khusra* house.

"I ran away from White Flower because she beat me too much and was angry when I didn't earn much money," she explains.

"She was like a servant in that place," one of her new friends adds. "They said, 'Do this, do that, get the drinks, turn on the tape.'"

Tasneem's new acquaintances are called Shaheen and Malika. Malika owns the house: she's a *khusra* of means. Shaheen is a friend who is visiting for the day. She has a feminine grace, and from certain angles her face looks attractive. Her mannerisms and body language have been refined to the point that, in some lights, she might pass for a woman, not just a man in drag. Such talents have their rewards—gold bangles and jewels sparkle on her wrists and ears. Tasneem has only a couple of plastic bangles and a ring with half the stones missing.

Shaheen and Malika want to talk about bras. They show me theirs: black with thick straps, lots of padding, and covered in hard scratchy lace that they describe as "English." They want to see mine, so I flash a Marks and Spencer strap at them to appreciative comments about the superb quality of yet more English lace. Tasneem doesn't join in. She squeals and holds on to her *kameez:* she doesn't want to be embarrassed by the state of her underwear.

Business is bad for Tasneem. We've gone back to the PCO. She's calling her regular customers to see when they are going to make their next visit. We've been here for half an hour. She bites her lips and furrows her brow as the phone rings. If someone answers she puts on a gently melodious, high-pitched voice. It doesn't work: nobody promises to visit. The man running the shop laughs and says something I don't catch. Tasneem jumps. It was an insult, and she barks back, dropping the girlie whine.

Shaheen stands behind us for the last ten minutes of the telesales admiring herself in the mirrored walls, tossing her hair around, pursing her lips, and adopting sexy poses. A handsome young man is sitting on the bench waiting for a turn on the phone. He is watching Shaheen go through her routine. Shaheen licks her lips and gives a little shiver. The man smiles—and not because he thinks she's amusing.

Shopping in Lahore with a group of *khusras* is a disconcerting experience: they are objects of silent scorn and hilarity, and I fit into no category except that of the freakish outsider. Tasneem and I walk through Heera Mandi hand in hand. Shaheen and Malika have joined us, *dupatta*-less and very, very brazen, with their fuchsia lipstick and gold platform sandals. Several extremely young clients approach them. The two *khusras* flirt with the boys and arrange to meet them after they've finished shopping. Tasneem has her *dupatta* drawn like a skin around her head, and she's glaring at everyone except me. "You're *my* friend," she hisses as she drags me across the road and away from Shaheen.

A five-minute walk takes you well outside Heera Mandi to more respectable parts of the Walled City. The crowds are thick in the bazaars and many people have looked a second or third time at the odd spectacle of *khusras* and a *goree*, a white woman, buying false eyelashes and glitter-speckled headbands. Tasneem continues to hiss, and she squeezes my hand whenever she sees an attractive man. She spots one she describes as "lovely" sitting outside a shoe shop, and she sighs. He's thin, with long straggly hair and a stretched-out, sleepy face.

We stand in a little pool of afternoon sunlight discussing the pros and cons of hair conditioners. The sun is shining on the unplucked parts of the *khusras'* beards, and on their chest hair, all bleached to resemble straw or a haze of yellow fluff. It's like a beautiful, golden aura. I'd say how pretty it looks, but they wouldn't like any mention of body or facial hair. They're girls, so they don't have any.

The Well of Death

Tasneem has invited me to see her dance in public. I arrive at Malika's house at half past seven with a big bottle of ice-cold beer and some potato chips. Tasneem likes beer and I've bought a nice one for her—Murrey's Classic. Malika says Tasneem isn't here: she's with an old *tamash been* and it might take some time. I sit in the corner and watch the others get ready for the dancing. There's plenty to see. Malika and another *khusra* called Razia are attending to their toilette. Razia has a lot of work to do: she has atrocious acne.

Malika sits on the mattress next to me so that I can observe the art of makeup. First there's the panstick—a light pinky-brown solid cream that's rubbed on the skin. A wet sponge smoothes it out and then another layer is added and smoothed. The sponge is then dampened and patted all over the face and the neck so that the pancake forms a second skin. The paste is set with talcum powder into a stiff but perfect mask. Next come the eyebrows — two thick black lines — and lots of red eye shadow. Malika brings out a selection of contact lenses and ponders over which she should use: blue, gold, or white stars? She chooses the white stars and puts them in with lots of blinking and tears that smudge the pancake. She turns toward me, waiting for the compliment, and I try to sound positive — but she looks like a terrifying mannequin: a person with doll's eyes.

Tasneem rushes in flapping a towel and sits next to me in a great panic. She hugs me, and I ask her if she wants the beer. Malika shouts at her to hurry up, so she runs to the bathhouse while the others apply thick black eyeliner and feathery, black-winged false eyelashes. The beer is put in an enormous new fridge. Inside are a couple of old samosas and a bottle of water.

Tricks with blusher and sparkly highlighter follow. Then comes lip liner, which is very dark, and lipstick, which is very red. It's topped with a slick of golden glitter. Malika's brother sits next to me and admires the effect. He's a rickshaw driver, about 40 years old, and quite good-looking.

Tasneem bursts back into the room and scrambles around, while Malika shouts about the time. Tasneem's panic is almost paralyzing. Razia, Malika, and her brother laugh as Tasneem rushes to get ready. The laughs are not nice, and Tasneem shrinks at the sound.

Tasneem's bottle of beer is taken out of the fridge, and Malika asks me if I want some. I say no, so she drinks it herself — as if it's for her. And, I suppose, it is in a way because Tasneem has no power in this household. She has no home of her own. She finds a place to live and work by playing the servant and the fool, for some-one else. So what belongs to Tasneem belongs to Malika. Tasneem asks me not to object. I shouldn't say anything to make Malika an-gry. Her face alternates between tense fright when she thinks the others cannot see and big smiles when she speaks to Malika.

Malika finishes off her makeup by reopening her makeup box and rummaging through the contents. She takes out a pot of glitter eye shadow that I gave to Tasneem last summer and spreads it over her eyelids. "It's very nice," she says. Tasneem nods.

Malika's brother announces that his sister isn't like the other *khusras*. "God didn't give her a *lun*. God didn't want her to be a man. God didn't want her to be a woman either."

Malika is in the cupboard that doubles as a changing room, and she invites me in to verify his story. There are piles of clothes everywhere and I have to squeeze past a mountain of *shalwaar kameez*. Malika is naked and proud of her body. She has no penis, nor any breasts, but her body shape is that of a man. I think she was born biologically male and has had her penis and testicles removed.

Tasneem looks troubled. "Don't tell anyone about her. It's a secret," she whispers, but it's Tasneem rather than Malika who wants to keep it a secret. A penisless Malika is tough competition—and she's also more like the *khusra* Tasneem wants to be. Tasneem is burning with envy.

A loud crashing comes from the little landing outside the room. A youth struggles to get past the goat tethered on the stairs and slips on some droppings. A tray with snacks and bottles of Coke smashes onto the concrete floor, and Malika begins a tirade. The youth shovels up the debris and the droppings, leaving it in a bubbling pile in the corner. He returns a few minutes later and bounces around the room distributing drinks and packs of *namkeen*. He's a slightly built young man of 18 or 19. He springs about in an uncoordinated manner on rubbery limbs and has a terrible, uncorrected harelip that never stops him from beaming at everyone. His hair is cut in a shiny, red-tinged bob that he flicks back with panache, and he wears men's Western-style clothes: a tweed jacket and cream-colored jeans.

The other *khusras* struggle into Western clothes too—girls' outfits: tight trousers, platform shoes, and stretch tops through which I can see lumps of English lace. They look like men in women's clothes: they have thick waists, flat bottoms, and skinny legs.

Malika's acned friend has done a miraculous job on her skin, and she teases Tasneem mercilessly as she jumps around the room pulling on her clothes. I can see the tears in Tasneem's eyes. Then the five of us—myself, Tasneem, Malika, the spotty one, and the one with the harelip—all pile into Malika's brother's rickshaw and leave for the circus. Tasneem is holding my hand and trying to talk, but the noise of the struggling engine drowns out all sound.

The circus is on the edge of Lahore: a shamble of tents erected inside a ragged wall of awnings and lit by fairy lights. This traveling show moves around the country spending a few weeks here and a few weeks there. We're met by "Security"—former soldiers who look as if they've fallen on very hard times—and we're ushered through the turnstile. Inside the fabric compound are several hundred men. I'm the only woman. A couple of games have attracted big crowds: one is a form of darts and the other is a large glass box from which the men are fishing for colored balls. At the far end of the compound is the Well of Death. It's surrounded by high scaffolding on which there are gaudy paintings of women with laughing faces, large breasts, tight suits, and dangling jewelry. I'm convinced that this circus has never seen anyone remotely like the women in these paintings.

The Well of Death is a common attraction in Pakistani circuses. This one is an ancient wooden structure about thirty feet high. Visitors climb steep steps to a viewing platform running around the rim. The Well is aptly named—it is a giant dark pit, and a couple of naked lightbulbs have been wrapped around a metal pole in the middle. A hairbrush dangles from a loop in one of the wires.

A door at the bottom lets the performers into the Well and a motorbike without a silencer roars in and parks in the middle. The *khusras* swing in after it and I wave to my four friends. There are ten other *khusras* with them. Most are in the same kind of girlie Western outfits, and all are preening madly. Tasneem is gesturing frantically. She's asking, "Who is the prettiest?" Malika asks the same question . . . three or four times.

It's bitterly cold and I'm shivering in my coat. The *khusras* must be freezing in their skimpy tops. Their breath, lit by the bulbs, rises in frosty clouds, and they start to dance to some distorted music. The routine is less classical South Asian dance and more Western disco, with a good deal of hip grinding and shaking of the stuffed bras. Some are good at it, some proficient, and a few are barely competent. The boy with the harelip is doing a strange dance, bouncing around on one leg with the other held in the air. Another is only 14 or 15, with a boy's haircut, men's jeans, a yellow polo shirt, and flip-flops. He doesn't have a padded bra, but he's wearing lots of badly applied makeup and is doing a stiff-legged skip between the other dancers. Two of the *khusras* look like women. Malika's brother says that they're like Malika—they were born without a penis. Judging by the curve of their bodies and the delicacy of their faces, he may be right.

Around seventy men hang over the rim watching the *khusras* dancing thirty feet below. They are ordinary men: laborers, tradesmen, and shopkeepers. They vary in age from their teens to their sixties. Some are throwing money down to the dancers. A smartly dressed middle-aged man with a big gold watch covers a particularly pretty *khusra* with a shower of ten-rupee [twenty-cent] notes. It's like confetti fluttering around the performers. The *khusra* blows him a kiss and scrambles onto the floor to pick up the notes before anyone else can claim them as her own. The sight of so much money has spurred on the dancers, and they stop every minute or so to use the hairbrush dangling on the light stand. The audience is agog.

The motorbike rider—the highlight of the show—enters the well. He revs up the bike and the *khusras* leave. He accelerates, climbing the walls of the Well, faster and faster, so that he's only a few feet from the rim. Only three men have stayed to watch. The rest have followed the *khusras*, who are now standing outside the Well doing another little dance and arranging appointments with the men. The bike act ends, the *khusras* go back into the Well, and the audience reassembles on the rim. It's a cycle that is repeated over and over again. Tasneem squeezes me during one of the intervals: she's negotiated a deal with one of the *tamash been* and is going to a hotel for the night after the show. Malika is also pleased: business is good.

"They're not girls," two customers tell me. "They're half-man, half-woman."

Presents

A tiny plastic Christmas tree has been put in the back window of Iqbal's restaurant. I've helped Tariq and the other sweepers to hang a dozen baubles and I'm feeling homesick. I've written cards and have a few presents to deliver. Jamila, the first person I intend to visit, is not at home: her husband says that she's gone away for a day or two to do some begging. Tasneem isn't at home either, but Malika tells me to sit and wait. She opens Tasneem's Christmas card and looks at the big fat Santa. "Oh," she laughs. "He looks like one of my husbands."

She unlocks a cupboard, pulls out a box, and settles herself next to me. She takes her jewelry out of the box piece by piece and spreads it on the mattress. "These cost fifty thousand rupees [$842]," she explains clanking half a dozen prettily worked bracelets. "They were from my second husband . . . a very nice husband." She dangles a necklace in front of my eyes. "This cost eight thousand rupees [$135]. It was from my fourth husband."

The donor of a pair of heavy gold earrings comes in and sits next to us. "He's my husband," Malika remarks, without identifying his number. The man is about 40. He's fat with a very round face and small beady eyes and speaks with a high-pitched, squeaky rasp that is strangely out of keeping with his substantial frame. He leaves his expensive, beautifully polished shoes by the door; his feet are smooth and soft. He gives Malika a present—a new ring. She squeals and makes a fuss, and then he leaves. "He's a nice husband," Malika comments and she puts the ring in the box with her other gold.

"Is he married?" I ask.

"Of course," Malika laughs. "All my husbands are married to women, and they come to me for fun."

Tasneem bursts into the room, upset and jittery. She's thrilled with the Christmas card, but I do note that she's looking at it upside down. It's understandable that she makes a mistake over the words: she's illiterate, and Pakistani books and cards open in the opposite way to English ones. Still, the card is placed on the cupboard with Father Christmas standing on his head.

Tasneem rummages through her suitcase and pulls out her two best dancing outfits — shimmering polyester *shalwaar kameez,* one in purple and one in orange. "Which one would you like?" she asks. "Which color do you like best?" I explain that I have lots of clothes and don't need any more, but she's persistent.

"But I want you to have one. Then we can be sisters. You gave me a suit in the hot weather, remember?"

Reluctantly I choose the purple one and Tasneem and I go to try on the outfits in the cupboard. Fortunately the purple suit is too small, so I say that I can't wear it.

"No, no," Tasneem cries. "We can go to the *darzi.*" She pulls me out of the cupboard and we head off to the tailor's. We walk a little way down the street and up into a big, well-decorated house. A group of *khusras* is standing in the doorway and they look at me as we pass by.

"Is it a boy?" they ask Tasneem.

"No, it's a girl," she answers.

"Really?" comes the reply.

I'm offended — people at home used to say that I was something of a beauty.

The building is an important *khusra* house. Two sit on the floor sewing. One of them measures my chest so that he can make the alterations to the suit. The *khusras* watch and I catch some of their whispers. "She could be a boy. She's got small tits."

I shout at them and they laugh.

A tall, well-built man in his late forties walks in with a proprietorial air and surveys the scene with confidence. He's wearing men's clothes and his hair is plastered in henna. He looks at me intently for a few uncomfortable seconds and then smiles.

"This is our guru," Tasneem explains. "He's very, very important. Before we do anything we have to ask his permission. I go like this." She walks over to the man, kneels down before him and kisses his feet. The guru smiles in a benevolent way and then leaves the room trailed by half a dozen acolytes. He's the head *khusra* in Heera Mandi.

The dress is ready in minutes and it fits. Tasneem takes me back to Malika's house and watches me change into the purple creation. She brushes my hair, ties it up, and arranges a sequinned *dupatta* over my head. Her eyes are full of tears and she hugs me.

"Now we're sisters," she snuffles.

Ramzan

At six o'clock on a cool, dark December evening boys are leaping from *charpoy* to *charpoy* in the courtyard while the young men near the pimps' den sing and play the *tabla*. A group of elderly women sit around a charcoal burner toasting their feet, smoking, and laughing, happy because it's Ramzan and they've just finished eating.

Ramzan, or Ramadan, is the Islamic month of fasting, and most Muslims from the religious mainstream fast from sunrise to sunset. Life in Pakistan moves at a slow pace: things happen without haste, if at all. During Ramzan, the country winds down even more. Offices and shops close early, restaurants are shut in the day, and people disappear during daylight hours. Some are asleep. They've reversed day and night.

Not everyone fasts, but it's good to be seen to be abstinent, and there is social and political pressure to comply. The ill, the pregnant, and the very young are excused. As a non-Muslim I'm excused too, but it seems insensitive to eat or drink in front of the hungry, so I shop furtively and eat quickly and in private. So do a few of those participating in the fast. In lots of offices there is a cupboard or little room scattered with crumbs, cigarette butts, and empty teacups.

Ramzan nights are for eating and playing cricket, though the minds of the religious are supposed to be focused upon spiritual reflection, not upon the excitements of the flesh. Sex is frowned upon during this time, so trade isn't good in Heera Mandi, and those prostitutes with a reputation to preserve will not be selling their services in the bazaar. It would be an insult to God. The poorest cannot afford religious scruples. In Tibbi Gali, Nazia is still on her doorstep.

The courtyard is eerily quiet in the middle of the night, partly because it is Ramzan and partly because it is so achingly cold. The popcorn and ice-cream *wallas* are taking a holiday but the *paan wallas* still come on their rounds. There must have been a low-key party next door because they bought a surprising quantity of sweet *paan* around midnight. A metal dish was lowered from the fourth floor on a rope, and the goods were transferred with a loud clanking as the dish bumped and scraped its way back up the walls.

A siren wakes me at half past three in the morning. It is the official wakeup call, given so that people can be sure to cook and eat before dawn breaks and the fast begins. The sweepers are out shortly after the siren sounds. Three figures dressed in old clothes bend over their brushes, their free hands alternately resting on their backs and swinging to the ground. Their heads are bandaged in ancient rags and long cloths hang over their faces. They use traditional *jharu,* brooms made of bundles of long sticks tied together at one end, and the rhythmic scratching of their brushes is a familiar sound in the courtyard in the hour immediately before and after dawn.

The Badshahi Masjid has been transformed into something resembling a Disney spectacular. It could be a scene from *Aladdin,* with a few flying carpets hovering around the minarets and whooshing between the domes. Thousands of fairy lights run in delicate streams over the walls and minarets and bright blue and green fluorescent strip lights illuminate the giant domes. These colored lights have been borrowed from the front of a nearby hotel and placed on the mosque for the last few days of Ramzan. The owner of this hotel has also paid for the fairy lights. He can afford this pretty offering because he is said to operate one of the biggest drug rings in the *mohalla.*

Ramzan has interrupted play on the grass. It's far too holy a time for frivolities like sport to be tolerated so near to the *masjid.* The addicts who usually live on the fringes by the trees have now taken up residence on the other side of the wall surrounding the field. When I passed by this afternoon they were huddled in little groups, smoking heroin. A couple of young men sat watching an older addict fill a syringe with pharmaceutical drugs and inject himself before they took their own turn. They were too engrossed to notice me. Another of them died last night. His body was in the bazaar this morning. It lay, with a collecting bowl at its feet, on the *charpoy* that is reserved for those who die without money or family to bury them. He looked desiccated and his skin had the sheen and color of the dates we eat to break our fast. There are new bodies on that *charpoy* every week.

Mullah Ali's Tazia

The twentieth of Ramzan is an important day in the Shia Muslim calendar: it is the death anniversary of Ali, cousin and son-in-law of the prophet Mohammed. Maha has told me that his body is being taken in a procession from Heera Mandi to a place just outside Bhati Gate. It is very unlikely that any of Ali's remains will be carried through the streets of Lahore, but the women of the *mohalla* seem to think that they will and I guess that is what matters. I'm informed that I must be ready at three-thirty. I must wear black. I must have no trace of makeup on my face, and I must not smile — it's not appropriate. It is a time for mourning. "We," several local women insist, "are Shias." Mourning is what Shias do best. They've made it into an art and a way of life.

We're passing from Heera Mandi Chowk to Tarranum Chowk, and above us hundreds of women are gathering to watch the procession from the second-, third-, and fourth-floor windows and balconies. I pause to look. It's breathtaking: the invisible prostitutes of the *mohalla* have come out from between their four walls.

Mullah Ali's remains — virtual or real — are encased in a large, heavily decorated silver casket. It is a *tazia* — a model of his tomb. There's a dome on the top, and people are throwing garlands over it so that it's quickly smothered in flowers. When the wind blows in our direction it carries a beautiful, sweet fragrance.

The men jostle and fight to join the teams carrying the *tazia* on their shoulders. Competition is so intense that the holy relic veers from one side of the street to another. It jerks up and down and is never level. They look as if they're going to drop the thing and its procession is anything but stately. It is a chaotic and at times bad-tempered cavalcade.

Young men stomp along ahead of Ali's *tazia*. Their arms are raised and there's a hollow thump as they beat their chests in unison. The women walk behind, corralled by a chain of men linking hands. I don't think it's an effort to keep the women away from the *tazia*. It seems more like a measure taken for their protection. The men are boisterous, and a very fine line divides their energy from destructive violence.

Tiny babies are passed toward the *tazia* over the men's heads. The lucky ones get their little bodies rubbed against the *tazia* or their heads pressed against the silver. None of them appreciate the spiritual significance of the act and all are bright red and screaming: their cries drowned in a sea of deafening appeals to God and to Ali.

We move on, bodies pressing from all sides. Hundreds of shoes litter the streets. The press is so tight and the speed of the crowd so erratic that it's easy to lose your footwear, especially when most people wear cheap plastic sandals or flip-flops that rarely fit. A policeman's rifle keeps sticking into my back and an obese, middle-aged woman grumbles because I'm moving too slowly. I tell her I can't move faster than the person in front of me, but she mutters something nasty in my face with thick, fetid breath. Maha interrupts her prayers and shouts that she's a dirty old bitch.

On the other side of Bhati Gate the procession spills out from the inner city into the wide space of condoned-off roads. Armored personnel carriers and policemen with guns watch the *tazia* veering through the crowds. An ambulance revs up and criss-crosses through the procession trying to make work for itself by mowing down children, old folk, and those with slow reaction time. And then, for a few seconds, no one is watching the *tazia*. The siren has sounded to end the fast, and everyone is looking at the food they've collected in their hands.

The ambulance clanks past again and people jump out of the way. This time it has a real mission: some of the men have beaten themselves unconscious with blades strung on long chains. Two of them are bleeding profusely from the multiple wounds they have sliced into their backs. Maha sighs in awe and is moved to tears by this proof of the men's religious devotion.

AIDS

Tibbi Gali is quiet in Ramzan. There are fewer customers, and although many of the women are still working, they've moved further into the shadow of their rooms. Nazia's madam sits with her thighs spread out over the step, but I can see nothing of the girl's face except the shimmer of her luminous makeup through the dark. Only the little group of women and girls who trade near the shop are clearly looking for business. They are huddling around a charcoal burner. Shela, the shopkeeper, is wearing woolly socks that have been toasted to a crisp by the fire. Sabina, the girl with the painful limp, is mixing a thick white fizzy paste on a plate. Shela looks intently at her reflection in the shard of a broken mirror and trowels the mixture onto her face.

"It's to make her white," Sabina explains. "It's bleach."

"Do you want some?" Shela offers me the plate. "It will be good for those dots on your face."

Shela sits in the center of the *gali* for twenty-five minutes while the bleach froths. She shouts at anyone who struggles to pass her. She's irritated because business is slow: there are five girls around the shop, and in half an hour only one has disappeared up a dark staircase with a young man. Sabina looks bored and leans against the wall to take the weight off her bad leg. She's always here in the shade, looking up the same narrow alley and talking to the same people about the same things. I've never once been to Tibbi Gali and *not* found her against the wall. This few meters of gray alleyway, Shela's shop, the narrow staircase, and her room are her whole world. She tells me she never leaves Tibbi Gali. "Why would I go outside?" she questions. "For what reason?"

A young woman gestures frantically from the house on the opposite side of the *gali* and offers me tea. The upstairs room of her house is clean and a lot of furniture is crammed into a very small space. Her younger sister, a child of about 12, is watching a soft-porn film that seems to be from the Middle East. A woman is doing a vulgar dance in beige Lycra leggings and the child is copying it, rolling around the floor doing pelvic thrusts and spreading her legs. The film is switched off and she sits down with a grin.

The two sisters and their mother are from Balochistan, Pakistan's most remote province. Their mother tongue is Balochi; they speak Urdu poorly and some Punjabi. The older sister says her brother got married fifteen months ago. She puts on the wedding video so she can point out key guests. She pauses the video so I can see her and another woman I know—but I barely recognize them. They've changed a lot since the wedding: both are shades of their former selves. They have each lost a dramatic amount of weight and the bloom in their skin has vanished. I can't tell whether it's tuberculosis or HIV or just general ill health, but there's something terribly wrong with these young women.

Condoms are a rarity in Heera Mandi. Everyone says the same thing: the clients don't use them. The junk and the rubbish dumped on the streets verifies it: a used condom is a rare, if unsavory, sight. The safe-sex message has not reached Heera Mandi, and there is no public or private awareness of the danger.

Sex and condoms are not compatible in the *mohalla*. Maha puts it this way: "The *tamash been* don't like condoms because it's not a natural feeling. They want a full feeling. Full sex. Condoms are not good because it's like putting their dick into a *shopper*"—a plastic bag.

Many women have never even heard of AIDS in Tibbi Gali, or, if they have, they think it's a disease of foreign homosexual men. Nazia's madam thinks of herself as an authority on the subject. She says, "You get AIDS from dirty men who do it in the ass." Higher-class women consider AIDS to be the exclusive preserve of *gandi* prostitutes in Tibbi Gali. Women with *izzat* would never dream of using condoms. Maha states with absolute conviction, "I can't get AIDS because I'm very neat and clean and all my husbands have been *sharif* men. My *kusi* is always fresh and lovely. Those *gandi kanjri* in Tibbi Gali will get AIDS because they are dirty and they have to do it the back way. They never wash their *kusi* and the *tamash been* in that place are dirty men with no money." She's wrong on so many of these points. Even the cheapest, oldest prostitute in Tibbi Gali washes herself in a bowl of water she keeps in the corner of her room.

Friday Prayers

The last Friday in Ramzan is especially holy and the Badshahi Masjid is packed. A separate, screened section has been set aside for the ladies but it's long since filled up, and the women are taking their places alongside the men in the quadrangle. I stand at the side with Maha's children and watch the crowds pour in. I recognize some of the men, and a few of the *khusras*, all of whom have elected to be in the men's section and not to fight for a place with the women. They are wearing men's clothes and have removed their makeup to show their respect.

We sit on the steps as thousands of worshippers form long lines in the quadrangle. Nisha is amazed. "So many people. It's like the whole world is here," she gasps.

When the prayers have finished, the worshippers stream out of the mosque and are assailed by the familiar army of limbless beggars. Another group is waiting for them too: lots of thin young men with straggly beards and green turbans are collecting on behalf of the Taliban. They get very little cash and most people ignore them.

Ariba's Boil

Ariba looks odd. I don't understand how such a small body can manufacture such a mammoth boil. She told me it was a burn that went bad, but it looks just like a giant pimple to me. It's yellow-green and the size of an acorn bulging beneath her nose. She needs to visit Dr. Qazi, but Maha hasn't noticed Ariba's embarrassment or her pain. She winces when she eats and talks. She has shown me the horrible thing several times and told me how much it hurts. These close viewings are testing my fondness for her, and I'm ashamed that I have to steel myself to stop recoiling from her worried, disfigured face.

Maha has missed the unmissable because she doesn't even see her children when they are standing right in front of her. And now I know why: I've just found half a dozen empty bottles of cough medicine beside the mattress. She's addicted to the stuff. It explains why she was so vile in Sewan Sharif: her supply had dried up before we even got there. It's Nena's job to collect the bottles and throw them away, but today she has failed in her duty, and Maha's habit is no longer a secret. She's on at least two bottles of Corex a day and her intake is increasing.

Corex is a popular cough medicine in Heera Mandi, and it dulls far more unhappiness than it ever treats bad chests. It's strong stuff—codeine, I think—and Maha buys it at a pharmacy near the Zakariya Hospital off the Ravi Road. The label advises adults to take two teaspoons three times a day. Maha swigs an entire bottle in seconds. She says it helps her to forget her loneliness and screen out the arguing and crying of the children. She confesses that she's been taking it for about six years, but she's upped her consumption recently to cope with Adnan's absence. Corex makes her drowsy, so she sleeps a lot. "I'd die if I couldn't take drugs," she states flatly.

Three pairs of large *ghungaroo* are stacked by the television. They're the traditional accessories worn by dancing girls; they cover the top of the ankles and reach halfway up the calves, a little like the leg pads worn by cricketers. They are very heavy because they are covered in tiny bells that tinkle and jangle with the slightest movement of their legs and feet. When dancers perform, the stamping of their feet and the sound of the *ghungaroo* become part of the music. These new *ghungaroo* belong to Maha, Nisha, and Nena. They're all planning to set up in business in the bazaar. They've identified a vacant *kotha,* they have the outfits, they've recruited the musicians, and they have enough desperate need.

"Nisha and Nena will sing and dance. That's all. No man is going to buy their *kusi,*" Maha declares angrily. "Swear to me, Louise. Swear to Shahbaz Qalandar. Make a promise to Mullah Ali. Look after your daughters. Tell your mother. Keep them safe. Keep them inside and safe. Don't let any man come to your house." Maha runs her fingers over the *ghungaroo* and grows quiet.

A fat elderly man eases himself through the door. He struggles with a stomach that has an existence independent of the rest of his body, and he's wearing a decorative *topi* (hat) that's embroidered with lots of miniature mirrors. His rings are even more impressive. The size of golf balls, they're the largest and least tasteful I've seen and he wears one on each wedge-shaped finger. It's a miracle he still has the use of his hands. He has come for the rent—and Maha doesn't have it. She explains that she's been fighting with her husband, and that he'll have to wait another two days until she can find some cash. For once God is on Maha's side: The siren sounds to announce that the day's fast is ending. The girls dash in with dates, a saucer of salt, and some *samosas* and the man eats as if he's been fasting continuously for a week. He leaves with a wave, a rub of his belly, and a promise to be back.

"He owns the building," Maha says. "Him and the old woman

downstairs. He's her husband. They've been together since they were young."

"Is he her proper husband?" I ask.

Maha's eyebrows lift. "Are you mad? She's a Heera Mandi woman. A *tawaif*. He's *sharif*"—noble and respectable—"and he's married to a woman from outside. But he still loves her. He still visits her and helps her. He collects the rent because she can't get up the stairs with her bad knees." Maha sucks in her cheeks and does a jerky, round-shouldered shuffle around the room. Her daughters laugh. "Strange," Maha says, half in fun and half in wonder, "he still loves the old *tawaif*."

Stars of Lollywood

A mellow Maha is huddled on the bed wrapped in a big fluffy blanket. She's had one and a half bottles of Corex and it's only two o'clock. A big plate of samosas dripping in oil lies on the mattress— Maha eats one and falls asleep. Nisha and Nena parade around the room showing me their hair and their latest dance routines. Nena is excited about her dancing career and twirls around refusing the samosas.

"Are you fasting?" I ask.

"No, I'm on a diet."

She's 13 and very, very slim.

"Look, I'm fat." She shows me her stomach, which is flat, and the girth of her hips, which is negligible. She has the yet-to-be-filled-out figure of an adolescent. Even before embarking on the diet, she didn't eat well. She has dark shadows under her eyes and a small, nasty cold sore. But, even with these ragged edges, she's still quietly, stylishly beautiful. Maha has taken her to the biggest promoter in Heera Mandi, and he's agreed that she has definite potential as a model or an actress, but he warned that the camera puts on pounds. I fall back in disbelief when she describes her weight problem. If Nena can put on a few pounds she'll look normal. If she loses any

she'll look like Nisha: a sunken-chested skeleton. She isn't convinced by my arguments and eats three chickpeas for the sake of her
career.

The female stars of the Pakistani film industry have traditionally
come from Heera Mandi. Respectable families wouldn't allow their
daughters to become actresses because acting, like dancing, was
synonymous with prostitution. Many Bollywood actresses have
murky pasts, and so did the actresses in Lollywood, the center of
the Pakistani movie industry, based in Lahore. Those women dancing, shimmying, and shaking across the screen in the Lollywood
films of the second half of the twentieth century were the girls from
Heera Mandi, known for their lascivious hip grinding and earthy
sensuality. A career in film and television no longer carries quite
such scandalous implications, but even today, a large proportion of
the country's film and soap stars were born and raised in Heera
Mandi. Here in the *mohalla* it's considered the best path to a better,
more respectable life.

During a visit to Karachi I met a remarkable woman who had
worked her way out of Heera Mandi. She is a film and television
star, around 50 and past her glory days, but her age is obscured by
well-toned skin, a rigorous maintenance regime, and a wickedly exciting enjoyment of life. She has a large apartment in a prosperous
suburb and is a gracious hostess; she keeps a well-appointed room
for entertaining guests, a well-stocked drinks cabinet, and a cook
who whips up delightful delicacies: tiny, plump samosas; miniature
kofte—meatballs served with a fiery lime pickle; little pastries sprinkled with toasted sesame seeds; and delicate, aromatic sweets laced
with cardamom, ground almonds, and chopped pistachio. On my
guided tour of the apartment I was shown a room carpeted in silk
Persian rugs and plush, embroidered cushions, the mirrored walls
reflecting a studied opulence. It was the star's private, very exclusive *kotha*.

A couple of other guests hung on the star's every word: men
enchanted by her sophistication and by a hard-to-define quality,
something instilled by a childhood spent in Heera Mandi. She was

well dressed—and well covered—but her *shalwaar kameez* was expertly tailored to reveal every curve of her figure, and she moved with a controlled but blatant sexuality: joking with the men, moving a little too quickly, kicking off her satin slippers, wriggling with laughter, aware of her body and its impact upon her guests. A butler appeared with another round of drinks, ice cubes clinking in crystal glasses on a silver tray. We gossiped and the star told tales of the old Heera Mandi as two stunningly handsome young men in white *shalwaar kameez* sat at her feet. She carried on talking, pulling up her *shalwaar* so that they could drizzle fragrant oils over her skin and begin to massage her calves. She shivered for a moment with pleasure and then resumed her conversation. For once, I wished I too could be a dancing girl.

The women of Heera Mandi are alive to this kind of dream: they fantasize about it but they know that only the luckiest women will build a successful career in Lollywood. Nisha, Nena, and Ariba keep well-thumbed film cards of famous actresses in a big box. They know everyone: where they came from, whom they loved, and whom they married. It's a life they aspire to and one that, for Nena, is tantalizingly within reach.

A Trip to Babar Market

I rarely leave the house with a bag these days. I manage with what the women of Heera Mandi have encouraged me to use for the sake of security: it's a "Pakistani pocket," otherwise known as a bra. If you wear one with a big enough cup size, you can fit just about anything in there. I take tissues, money, the key to my room, a contact lens case and a little bottle of contact lens fluid, my sketch map of the Walled City, and sometimes my mobile phone. It's fortunate I wear a *dupatta*, otherwise I would look as if I have immense and very lumpy breasts.

Last night, though, I went out with my handbag. I needed it to hide the vodka I was taking to a party given by a woman who lives near Iqbal, and not even the most generous of bras could accommodate a half-liter bottle. After the party I called in at Maha's house to take Ariba some more ointment for her boil. I lolled around on the bed eating *paan* in an unconvincing effort to fit in with the local scene. It's like eating a mouthful of spiced grit. Maha flopped on the mattress with me after downing another bottle of Corex.

Ariba sat on the sofa looking shifty-eyed, with that all-too-fulsome grin I remembered from her previous scramblings through my belongings. My bag lay next to her but was hidden from my view. I did a mental check through its contents. There was no money for her to steal, so I calculated that she could safely get on with the hunt. It was only when I got home that I remembered my beloved fountain pen. Today, no one has any idea where it is. Ariba gives me a vacant smile and shakes her head when I ask if she's seen it.

I speak quietly to Ariba, who is sitting in a filthy dress on one of the steps to the grand and long-locked doors of the courtyard's old houses, her hair dangling in her eyes. I offer her a reward if she can find my pen: I will take her to Babar Market and buy her another new suit. She nods and says she'll try. An hour later I'm sitting on her mother's balcony. Ariba has thought more about the offer and comes to speak to me. Yes, she says, she will scour the house for the pen tonight and, she adds, "Can we go to Babar Market tomorrow afternoon? And, as well as the suit, I also need a new pair of shoes."

The first thing I see this morning when I open the window is Ariba standing in the courtyard with a great grin on her face. She's waving my pen at me. I'm happy to see the pen, but I'm also caught in a dilemma. I promised to take Ariba to Babar Market if she could find it, but I feel hurt that she stole from me. So I've decided to buy all Maha's children a new outfit. In this way Ariba gets her new suit without it seeming like a reward for dishonesty.

We leave in a rickshaw once the pen is safely back in my room.

Maha has stayed at home, too drowsy from the Corex to walk a straight line or speak without blurring her words. Babar Market is packed with women shopping in preparation for Eid. The ceilings of the shops and the spaces between stalls are festooned with tinsel and fairy lights, and the children stand in a huddle and gaze at the colors. Standing back from them for a moment, I see them properly and in context for the first time. They're a wretched little bunch. Nisha looks like a patient at the TB hospital—the place she really should be. Ariba is utterly filthy, with crawling, matted hair, and she's thrown out of shop after shop when the assistants see how the months of dirt have worked into the weave of her clothes. Sofiya is also dirty, her face caked in snot, her hair hanging to her nose. She's not wearing any shoes. I'm escorting a group of children who look like street urchins—only the street urchins would be better behaved.

Mutazar is at his worst. The 5-year-old boy has the face of an angel but behaves like a devil. While we stand discussing the merits of fabrics, colors, and sizes of pretty suits, Mutazar rams our legs with his head. He screams and shouts and whines. And now he has found a better way to attract attention. One of the fairy lights surrounding the door to a shop has broken and Mutazar retrieves a piece of broken glass from the floor. He uses it to carve a deep and bloody channel in Nisha's foot.

The girls are thrilled with their clothes. Nena has a modern, stylish denim *shalwaar kameez:* it looks like a pair of jeans and a long fitted tunic. Nisha has a "flapper"—a shorter tunic with bell-bottom trousers—in lush green velvet. And Ariba, radiant and clutching a plastic bag to her chest in the back of the rickshaw, has chosen a startling party-time *shalwaar kameez* in luminous pink with a lace and sequin *dupatta.*

At home Maha is a little more alert. She's wearing a tracksuit and Nena is embarrassed because you can see her mother's legs. Maha doesn't notice her daughter's discomfort. There's only one thing on her mind: "Do I look fat?" she asks, doing a little jig on the spot.

I lie. Again. And I think of the cycles of unhappiness emanating from Maha's misery: cycles that are eating up her children and will probably consume their children in turn.

"They Saw Me Dance . . . and They Died.*"*

This is a landmark evening: Ariba has been scrubbed, and her hair washed and combed. Her boil is beginning to heal and she looks like a different girl. Maha is moving briskly between the rooms, shouting orders, stirring the pot balanced on the gas ring, plunging her hands into a bucket of suds to rub at stains on the soaking clothes, and berating her daughters for their failings. Everyone is in trouble. Maha is trying to give up Corex and we are all suffering with her. It reminds me of Sewan Sharif.

Maha is raging about the absent Adnan. "He only comes to see me for *kusi,* and if he doesn't have *kusi* he doesn't give me any money." She says it was the same after she gave birth to Mutazaar and Adnan elected not to visit. She had to go back into the business when he was only a few weeks old. Maha says it was hard, but there was no alternative. She worked in Defence, an expensive suburb, and left the children at home by themselves. The madam charged very high prices and kept half the fee herself. Her best girls could command around twenty thousand rupees a night—about $335. She lived well on this: she had beautiful clothes, a gorgeous house, and three cars. No one knew she was a madam because she was the charming wife of an army major.

Nisha is sulking on the mattress, her face set into a pinched, thin-lipped grimace. Maha waves the cooking spoon at me. "She won't wear the *ghungaroo.* The little *kusi* won't even put them on."

Nisha huffs into her *dupatta.* The now shining and glossy Ariba stands by the door. It's her usual stance—only half in the room—so she can make a quick getaway if things seem tense. "She's a *gandi* girl," Maha shrieks and points at Ariba. "She stank. I had to clean her with a brush. People in the bazaar are saying my daughter is *gandi."* Ariba turns on her heels and flees.

Nena is, as usual, taking the route of compromise and cooperation, and is trying to score points by cleaning. She's crouching down, moving along the edges of the room, brushing up bits of old food. She stops to put on a tape of Celine Dion's "My Heart Will Go On." This is the girls' current favorite song, but for some reason Maha has taken against it tonight.

She erupts with an enormous shout. "Switch it off. Get on and help me. When I was your age I was in the bazaar and dancing for my food and my clothes. I had lots and lots of men. I danced every night and I had so much money. I was rich. The men, they saw me dance . . . and they *died*."

Nena carries on brushing, never lifting her eyes from the floor.

Christmas

The Christmas service was supposed to start at half past ten. This was optimistic: when I opened the door at twenty-five past the pastor was sitting alone in his church dressed in his finest Western suit and shiniest shoes. An hour later the congregation is still trickling in, laughing and waving at their friends and relatives. Tariq and his family arrive toward the end of the service and launch into the singing. Tariq is unstinting with the tambourine and with his smiles, and we leave afterward in a jolly, tightly packed group, the women urging me to keep close so that no stranger can come near me.

Tariq and the other Christians who work in Heera Mandi live on the edge of the red-light area in a compound that's entered through a beautifully carved wooden gate. It's been there for centuries — the entrance to a grand *haveli*. It must have been glorious decades ago, but it has long since fallen into a shambles of powdery bricks and rotting wood: a ramshackle mess inhabited by dozens of families.

At street level the front of what is left of the *haveli* is divided into lots of small shops selling shoes, food, and spare parts. Up above the din and the chaos of the street, fifty yards of perfectly proportioned, exquisitely carved windows, balconies, and trellises run the length of the building. Most are piled with the residents' possessions and drying clothes. Above the balconies, yet more elaborate windows are crowned by a cornice of bricks and fancy plasterwork that, long ago, must have been impressive but now looks as if it's about to topple into the street.

Tiny, single-storey brick houses have been built inside the courtyard of the *haveli*. Most have a single room and are in desperate need of repair. Narrow passageways run between the houses. Tariq's house is a low, three-roomed building that he shares with his parents, his wife, and their three children. The family has lived in this impoverished, scrupulously clean home for fifty years. Tariq's parents look ancient even though he himself is still a young man: the last of their seven children. His mother is crippled, and her daughters-in-law take turns massaging her legs to give her some relief from the pain. She laughs a lot and wants to know about my children and why I have only three.

Tariq's father is blind. He has milky-white eyes in a worn, wrinkled version of his son's gentle face. I like him even though he disconcerts me by belching with astonishing force. He's been blind for five years. Like his son, he was a sweeper and was employed to clean the sewers in the old city. One day he came across a pocket of chemicals while he was underground: they'd seeped, or been discarded, from a goldsmith's shop. They burnt his eyes totally and irrecoverably. In Pakistan, there is no insurance, no compensation, and no disability benefit. There is no one to help him but his sons.

He doesn't complain but sits smiling on a *charpoy*. He thinks life in Lahore is better for people like him nowadays. His grandfather came to the city from a village and he worked as a soldier for the British. Lahore was different then. "There weren't so many Muslims and I can remember, when I was a boy, that there was jungle around Lahore. Now there's nothing left," he adds with a sigh.

The sweepers want to hear about the differences between my home and Pakistan. They're most interested in the English refuse collection system, and they gather around to marvel, shaking their heads and looking from one to another in disbelief. I explain that English households gather up their rubbish and put it in bins, and then it's collected every week by the council. They're incredulous.

Tariq's wife and her sister-in-law are busy cooking in their best Christmas clothes. I'd help them but I've been asked to sing English Christmas carols. I've been practicing for days but it's a poor effort. It's hard to sing "Silent Night" or "Little Town of Bethlehem" to the accompaniment of a harmonium—especially when the musician, who is Tariq's very gifted brother—has no previous knowledge of the melody.

The old lady thinks it's good and gurgles through the performance. She carries on gurgling through the beautifully prepared meal, sucking bits of soft meat off the pieces of chicken before passing them to her sons, who can use their teeth on the meat still clinging to the bones. Tariq helps his father with his meal: adjusting his plate; scooping up the food that falls onto the *charpoy*, and wiping the rice from the old man's chin and clothes.

A succession of guests arrive at the house bringing little gifts for the family: a cake, a card, or a small toy for the children. They tell us about a beautiful crib in a house on the other side of the *haveli*, so after the meal, we leave the old people and go to look at this wondrous crib. The house belongs to an extended family with fifteen children. The room is tidy: there are plates in a wire rack, a frequently patched quilt on the bed, and photographs of Indian and Pakistani film stars have been cut out of magazines and stuck into plastic picture frames. The walls and the ceiling have been covered in rolls of white paper especially for Christmas. Tariq says it makes it new and pure and clean, like baby Jesus.

The crib is in the corner of the room, illuminated by a single naked lightbulb. The family has built a stable with cardboard boxes and filled it with handfuls of straw. The children crowd around enchanted by the Christmas scene. The pottery figures are arranged just so: Mary is gazing at her tiny son; the animals are dotted around; and the shepherds and the wise men are kneeling in adoration, each one still wrapped in a thick, slightly opaque plastic bag.

"Sister," Tariq smiles and shakes his head in amazement. "Isn't it lovely."

"Yes," I agree. There is no doubt.

Eid-al-Fitr

Maha's catering has been upset by the moon. The Islamic calendar isn't entirely predictable. Unlike the Western calendar, which is determined by the solar year, the Islamic calendar follows the lunar year. Eid-al-Fitr begins only when the new moon has been spotted, and although the astronomers have a fairly good idea when this will happen, there's a degree of flexibility. Last night could have been Chand Raat—Moon Night—but cloud cover meant the moon didn't appear, so today all of Maha's celebratory food is sitting in the fridge.

Tonight will be Chand Raat, the night before the biggest holiday in the Muslim calendar. In social terms it is the Islamic equivalent of Christmas Eve. The children are excited, all the women are cooking, presents are being wrapped, houses are being decorated, and the markets are busy with last-minute shoppers. Nisha has covered my feet and hands with *mehindi*—henna decorations, especially for Eid. The combination of Nisha's lack of expertise and my pale skin has an unfortunate effect: patches of color that were supposed to be stars and flowers look like a virulent skin disease.

Babar Market is heaving. Fat matrons in black *burqas* are being particularly forceful in their use of the battering-ram technique to pass through the crowds. Maha haggles over some fake gold and diamonds and comes away pleased with the deal. In the bazaar we stop by the date and nut stall to buy great bags of pistachios, *namkeen,* dates, and peanuts. We're all so excited: we're going to have a feast.

Eid morning dawns warm and bright. The Badshahi Masjid is packed and the hundreds of worshippers who have been unable to enter the mosque are praying on the grass. Below Iqbal's house cars are parked haphazardly in the road creating an impassable jam. An expensive new Japanese saloon has been left without its handbrake on and has rolled into the side of another car. A four-wheel-drive vehicle is parked very precisely in the middle of the road so that only the slenderest of pedestrians can squeeze by.

The bits of Heera Mandi that are seen by the middle- and upper-class worshipers of Lahore this Eid have been purified so that they don't cause offense. The park and the streets have been swept and the addicts have been rounded up and shipped out of town. They've been dumped so far away from the *masjid* that it will take them days to find their way home. The authorities have organized an exceptionally vigorous bout of cleansing, and a lot of powdered lime has been scattered in the smelliest places. Even so, I'm not sure that the owners of those polished Hondas and Toyotas would be pleased to know they are parked in the customary urinal.

Balloon sellers, beggars, and ice-cream carts are poised around the entrances to the mosque, waiting for prayers to finish. From Iqbal's roof terrace we can see the lines of people in the quadrangle standing, kneeling, and bending in prayer: a broad field of light, bright colors. We can hear the prayers of the *mullahs* and a soft swish of sound as thousands of men, women, and children sink to their knees.

The fat clusters of balloons bob up and down as the worshippers pick up their prayer mats and begin to stream out of the mosque. Most people have come on foot, and they skirt the edges of the red-light area. They are ordinary people in their very finest clothes. Lots of the outfits are new. The fabrics are stiff with starch and their owners are self-consciously proud. The children have freshly washed hair and shiny faces. They're on their best behavior. This Eid day at the Badshahi Masjid is a joyous spectacle.

S hamsa lives in the corner of the courtyard. She didn't attend morning prayers; she didn't even wake up until midday. She stays with a Bangladeshi woman and her husband.

The woman is in her late thirties, with bad, broken teeth and weak eyes, and she's too old and thin to make money from the business herself. Her husband is a small, round man. His hair is speckled with gray and white at the temples, but he keeps his moustache blue-black with dye and lavishes attention on it so that it's clipped and brushed and glossed into rigid order. He devotes the same care to the gun that hangs from the ceiling in a polished leather holster.

The couple pimp girls from the house. One is the woman's daughter.

"I had four children," she explains looking at a pretty child playing in the room with a gang of others, "but this is the only one left. The rest died when they were babies."

The daughter is about 12 but very small and delicate. She has the body of a 9-year-old but a sexual precociousness that could make her pass for 30. She has full lips, wears a lot of makeup, and totters around on platform shoes that raise her to about five feet tall. It's disturbing to watch her play with the younger children because she moves so provocatively. She has been "dancing" for about a year and is the "baby" pimped by her father on the street corner. Not all the babies that the pimps offer are quite as young. They often try to pass off older girls and young women as younger than they really are. It's good for business to say a 20-year-old is actually only 13.

Shamsa is their other source of income. First they claim that she is the woman's sister, but half an hour later the story is different: she's their niece. At other times she's a friend. Despite what they say I'm sure she's not closely related to the couple.

She bursts out of the washroom, delirious with happiness: she's volcanic. It's Eid Day, she sings, and we're going for a walk.

"Where?" I ask

"The bazaar. Many, many places," she shrieks and twirls around the room. She runs to the window and, snatching a mirror from the madam, begins to apply another coat of makeup: thick eyeliner, pink lipstick and then, bizarrely, a shiny slab of dazzling, glittering green gel to her lips and eyes.

She drags me down the spiral stairwell, chatting incessantly about how happy she is. She's wearing dramatic jewelry, a flowery white-and-purple *shalwaar kameez,* and a *dupatta* draped over her shoulder. I shrink beside her as she links arms with me and talks about her rich men friends.

She stomps through the streets ignoring the looks of the men who are pausing to gape. For once no one is noticing me. They are looking at Shamsa—*dupatta*-less and in green glitter lip gloss. I'm seriously confused. She doesn't seem to hear the comments the men are making, and she hasn't noticed the way the bazaar is grinding to a halt around her. No women behave this way in Heera Mandi.

It is, I grant, a special day and everyone is wearing their best. Perhaps Shamsa is just excited by the carnival atmosphere in the *galis* and the bazaars. The restaurants are doing a roaring trade: the benches in the street are full, and the snack merchants and balloon sellers are busy with customers. Little girls dressed in brightly colored dresses edged with bits of nylon lace hold hands and look alternately thrilled and terrified. Shouting, giggling teenage girls have been let out wearing makeup and high-heeled shoes that stop them moving more than a few feet from their homes. Everyone wants a ride in a *tanga* and the lucky boys get to ride on a horse. Youths gallop bareback through the streets, and I'm sure someone is going to be trampled. Younger boys sit nervously on saddles and are led around at a trot.

We stop at the juice stall opposite the church and sit on stools in full view of the bazaar. It's the season for deliciously tart pomegranate juice, but it's hard to appreciate when you have an audience of rapt men. Shamsa is brazen, laughing with the juice *walla* and looking him right in the eye. He's having a good joke with her and she's enjoying it. I can hardly tell what she's saying. The conversation is hard work because she is speaking in Punjabi and when we talk in Urdu her accent is thick. She says she came from Multan—another important but smaller city in Punjab—but it's better here: the men are richer and there are more of them.

Now we're off to the fort and she wants me to get my camera. The fort is full of families and packs of youths on day trips. Today, the star attraction is not the exquisite Shisha Mahal—the Mirrored Palace—but Shamsa lying on the grass or leaning against a tree as if she's doing a photo shoot for a risqué Indian film. She's oblivious to the stampede of teenage boys and men around her.

As we walk back down to the fort's entrance she looks at me for a moment and says that I'm a *sharif* woman because I'm wearing my *dupatta*. She sounds almost wistful—and then forgets about it to marvel over three Western people who are looking at the board that gives information on the fort's history. We stand near them. They are British: two women and a man. The women are wearing *shalwaar kameez* but they don't have *dupattas*. There's nothing unusual about that. What Shamsa finds so stunning is that one of the women is black.

"Look. Look," she says in amused horror. "Look at the color. Look at the color. She's so black." She thinks it's funny and a bit sad.

We're off again leaving the black woman to suffer her way around the fort unaware of her deep misfortune. Shamsa is taking me for tea and sweets. We share a plate of *halva* in a tea shop, and she talks at me nonstop in Punjabi. She hasn't realized I understand only half of what she's saying. It doesn't seem to matter: she's really pleased. So is the owner of the tea shop: he says the snacks are on him and he'll see her later.

A feast is spread out on a white *chador* on the floor of Maha's best room: platters of chicken fried with tomatoes, chili, and ginger; bowls of mint and cucumber stirred into creamy yogurt; a great pile of freshly baked *rotiya* and *naan;* and a deep dish of tender lamb smothered in a rich, dark brown gravy. Maha is fussing around, excitedly handing out plates to guests and insisting that they should take yet another helping whether they're still hungry or not. All the family are in their best clothes—even Ariba looks presentable in a clean *shalwaar kameez*, rushing back and forth between the rooms with two of her cousins, the sons of Maha's dead sister. Maha grabs the boys whenever they pass by, squeezing them and pressing them to her bosom, her eyes awash with tears as she ruffles their hair. They live with their father, a man from outside Heera Mandi who, to the eternal horror of his official wife, took them in when their mother died. Maha says their stepmother is a bitch who treats the boys cruelly, sending them out on bitter winter evenings without a coat or any hot food in their bellies.

Visitors have been arriving all afternoon. Most are people poorer than Maha who have come to wish her Eid Mubarak—Happy Eid—and to pay their respects. The tailor is here with his brother; the Pathani woman from upstairs sits next to her children, fidgeting and looking impatient; and a couple of scrawny, dark-skinned sweeper ladies crouch by the door forever smiling. Maha gives them Eid presents: gifts of money and brightly colored glass bangles dusted with glitter. More visitors stream through the door and Maha rearranges her own gifts—the perfume and chocolates from me—on top of the fridge so that everyone can see they are from abroad. She's enjoying playing host, enjoying her little acts of kindness and the glee of the sweeper women as they leave with a hundred rupees. She catches her nephew's arm as he races by and presses a thousand rupees ($17) into his hand. "For warm coats for you and your brother," she instructs. This generosity will cost Maha dearly when she can't pay next month's rent, but for now she doesn't care about the future. She's happy in the moment.

"It's Eid," she laughs, "and I'm with my family and my sister's children."

For once, the Corex sits unopened behind the fridge.

Ankle Bells and Shia Blades

Muharram is the most important month in the Shia religious calendar and four large canopies stretch over the end of Fort Road, covering *sabils*, or water stalls. They're held in place by stakes and ropes that ensnare pedestrians and garrote men on motorbikes. Farther up the road, opposite the dump, lie other smaller shelters: filthy tattered sheets and blankets hooked over the railings, their edges weighed down with rocks. Bundles of rags are strewn along the length of the wall. Some of the bundles twitch, creating a buzz as the flies lift into the air. Others are immobile and surrounded by bits of animal—a chicken head, a goat hoof—dragged from the rubbish dump by the local dogs. Some of the piles of rags stink so badly that their contents— animal or human—must have died some time ago.

The picture outside Iqbal's house is cleaner and greener. The terra-cotta pots overflow with blooms. Every morning and late each afternoon one of the boys working in the restaurant waters them and sprinkles the road with water from a hose. It is early evening and the tiled floor of the roof terrace has also been dampened: for a brief time the water stills Lahore's fine powdered dust and cools the air as it evaporates.

The sun is a mellow orange: a crisp, perfect circle sinking in the gap between a minaret and the domes of the Badshahi Masjid. The bamboo scaffolding that has stood in the principal arch for months has gone, and if I shift my vision fractionally to screen out the addicts camped opposite the dump, I can see only perfection in the beauty of the mosque and the setting sun. From this angle there can be no better place to be.

The *azan* is called and the fast-disappearing sun tinges a few high clouds with pink. The canopies below us are being lit. Behind the mosque, far in the distance, the neon advertising signs have been switched on, blighting the timelessness of the view.

Dozens more *sabils* have been erected in the streets of the *mohalla*. They are distributing refreshments in memory of the thirst of the Shia martyrs. On the corner, the drug dealer's *sabil* is decked in a green awning, and in front of it are two giant shiny silver *panje*— Shia hands—cradled by the Islamic crescent. The ensemble is illuminated by regal chandeliers in patriotic green and white.

Almost immediately below me a canopy covers a grand *sabil*—more sophisticated than the drug dealer's because it serves its water with chunks of ice chipped from a mammoth glacier. The ice block and the water are set on a giant red cloth, and a fat man sits waiting to serve the pilgrims. Loud religious music blares out from a poor-quality tape deck; an emotional man sings about the martyred Hussain with the beat of the music provided not by drums but by the sound of men beating their chests. The *sabil* also boasts a picture gallery: a scene of a battlefield scattered with body parts, a rain of blood as one of the Shia martyrs is sliced into thousands of pieces by a whirlwind sword. A little way toward Roshnai Gate, the middle-aged woman from next door is offering snacks under her own faded canopy. She's gaining religious merit. Dressed in black and shoeless, she is suffering with the martyrs. It's drawn in the sadness of every line on her face.

*A*crowd is gathering around Iqbal as he sits on the roof terrace. He's signing official papers recognizing and so legitimizing the identities of those gathered around his table. He regularly acts as a referee, guarantor, and witness for many of the people of Heera Mandi because Iqbal is like them and yet he's not. He's from this community, but he is also a professor at the National Academy of Art and a nationally renowned artist. To the people who live beyond the *mohalla,* he will always be the tough, streetwise son of a courtesan who grew up in a brothel and carried a gun to college, but he's still recognized in a way that the rest of Heera Mandi never will be. He can put his name on a document and it will carry weight: he has an address, a title, and a career.

There are so many of these signing sessions in Iqbal's house because authentic documentation is hard to obtain in Heera Mandi. The children born to the women of the *mohalla* don't have fathers: there's no family name to put on birth certificates, passports, and identity cards because, in such a male-dominated society, children without fathers are not supposed to exist. When Heera Mandi's fatherless children encounter the bureaucratic world, Iqbal signs their papers to prove that these semiliterate and illiterate people are real.

"We are the same," he says to me as he waves them goodbye. He's looking over the railings into the busy street below and smiles with a wry resignation. "I don't know who my father was either. Perhaps he was a painter too."

The Shia and the Sunni

Islam is fractured into many groups and the most important divide is between the Sunni and Shia sects. Worldwide, the Shia are in a minority. Most live in Iraq, Yemen, Pakistan, Afghanistan, and in Iran, where they form the religious government of the ayatollahs. In Pakistan, only 10 to 20 percent of the population is Shia, and they believe themselves to be persecuted and discriminated against by the Sunni majority. Pakistan is wracked by sectarian and political conflict, and both Sunni and Shia terror groups operate in the country. Karachi is the scene of much bloodletting, but serious sectarian tension exists here in Punjab. In Pakistan, Islamic extremists fight the West and they fight their fellow Muslims too.

The Shia-Sunni rift opened shortly after the birth of Islam and has been unbridgeable ever since. The events surrounding the rift are replayed every year during Muharram, accentuating the sectarian divide and stirring the passions that sustain it. At the heart of the split was a dispute over who should lead Islam after the death of the prophet Mohammed. One group, who became known as the Sunnis, favored a nonhereditary head elected by scholars and community leaders. Another group argued that members of the Prophet's family should lead the faith. This group rallied around Ali, cousin and son-in-law of the Prophet; they became known as the partisans of Ali, or the Shia.

Ali eventually became caliph but was later assassinated. His sons Hasan and Hussain—grandsons of the Prophet through his daughter Fatima—were then forcibly prevented from succeeding their father. Hasan was poisoned, and Hussain set out from Mecca with his family and a small band of followers to overthrow the existing caliph, Yazid, a man who had a well-established reputation as a debauched tyrant. They failed. At Karbala, on the banks of the Euphrates in modern-day Iraq, they came under siege from Yazid's forces from the second to the tenth day of Muharram. The children, deprived of water in the terrible heat, were said to have cried in desperation while the army of the evil Yazid remained deaf to their suffering.

On the tenth of Muharram, an overwhelmingly larger force slaughtered Hussain and many of his followers. According to Shia history, Hussain's infant son was killed by an arrow and his surviving family was taken to Yazid's court, where the victor crowed over Hussain's severed head and beat it with a cane. While they were held captive in Yazid's palace another of Hussain's children died, and Zeyneb, his sister, held the first *majlis*, or lamentation assembly. These *majalis* continue to this day. The first ten days of Muharram, marking the anniversary of the siege and the battle, are the most important of the Shia calendar. In Heera Mandi they are days of *majalis*, processions, and heartbreaking gloom because most people living here are Shia.

The suffering and martyrdom of Karbala define the Shia. They believe themselves to be under attack, and every year they grieve for their martyred leaders. Part of this mourning is *matam*, an act of ritual lamentation that takes different forms: in Lahore the two most common forms are *hath ka matam*, which involves beating the chest with the open palms, and *zanjiri matam*, self-flagellation, in which the back is whipped with blades strung on metal chains.

Performing *matam* is a declaration of Shiism, reaffirming the community of the faithful and strengthening it against outside threats. Muharram remembrances among the Shias of Heera Mandi bind this community in a way few other things do. If a family wants to impress visitors, it shows them the scars on their men's backs. Relatives describe how much they bled and report that the only thing they put on the open wounds was oil. Possessing scars is like having a chest full of medals: proof of devotion and masculinity. As soon as a boy can walk he's presented with chains and blades with which to practice. Unlike the grown-up versions, the blades are blunt and made from lightweight aluminium or plastic. Families are proud to see their little boys lashing their backs and consider it a sad day when men grow old and stop performing *matam*. Iqbal said his family was disappointed and his wife a little irked when, at last, he hung up his flails.

The more orthodox Sunni majority in Pakistan look down upon such rituals as semibarbaric. Ashura, the Night of Murders commemorating the final struggle and deaths at Karbala, has frequently triggered tensions between Sunnis and Shias. Today, in the old city, there's fear that passion will spill over into clashes, so police have been deployed in every bazaar in which Muharram observances are held.

Black is the color of Muharram. The most religious of the Shia wear nothing but black for the first, sad ten days of the month, and prostitutes with any status to maintain will refuse to wear makeup. Half a dozen local men have already approached me to thank me for wearing black too. They say they appreciate the respect I'm showing to them and their martyrs.

Chains are important Muharram accessories. Maha says they're reminders of the siege at Karbala, the imprisonment by the vicious Yazid, and the suffering of all Shias. There are different designs on the same basic theme: a thick metal band is fastened around the neck. Several long chains are attached to it and reach down to the ankles where they are fastened by other metal rings. The ailing Nisha has thin, lightweight ones that are far too long and constantly trip her up. In contrast, strong, devout, and particularly macho young men drag around the bazaar in heavy, industrial-grade chains.

Shia and Sufi rituals structure the lives of traditional brothel families and provide a refuge and a source of comfort. Islam is something they can depend on and something that will be of greater personal significance as they age. They certainly seem to spend more time on religious activities as they approach 40; they carry prayer beads and pray more. They don't go to the mosque to do this but worship at home on their prayer mats. Women aren't usually allowed into the mosques at prayer time, either because of religious edict or social practice, but they can go in at other times providing they're veiled. The last time I visited the majestic Badshahi Masjid a group of women were there on a day trip from their village: they had brought their offspring—inquisitive toddlers and energetic children—together with a large bag containing a picnic lunch and blankets, which they spread on the steps leading to the glories of the central dome.

Sufis like Shahbaz Qalandar and Data Ganj Bakhsh figure prominently in the Shia Islam of Heera Mandi. Local women visit the
tombs of religious figures, taking offerings of food to be distributed to
the poor. Everyone appears to have an unswayable belief in charms
and amulets, and the members of the Prophet's family are loved to
such a degree that they are believed to possess divine powers. They
can intercede with God and help work miracles. The metal *panje*—the
mystical hands—that rise high above the Walled City are protective:
the five fingers representing the Prophet, Ali, Fatima, Hasan, and
Hussain. In Shia folklore the *tazia* that are carried during Muharram
processions are thought to be blessed, and Fatima, the Prophet's
daughter, attends every *majlis* in person.

The deeply felt religiosity of most women in Heera Mandi isn't
matched by a strong command of religious doctrine. In Shahi Mohalla the stories of Karbala and Shia suffering have assumed a life of
their own, and bear only a vague similarity to the records and inter-
pretations in scholarly books. I've sat through impassioned mono-
logues as women describe the vile tortures perpetrated upon the
martyrs. They pore over highly stylized pictures of the bloody Karbala battlefield and point out important individuals. They talk about
the martyrs as if they're well-loved family members whose suffer-
ings happened just recently, not thirteen hundred years ago. All give
careful attention to the trials of the martyrs' womenfolk. Maha
weeps as she recounts how they were stripped of their veils, but
bravely struggled to preserve their honor by covering their faces
with their hair. Shiism, the religion of a stigmatized sect that believes itself under siege, is particularly suited to Heera Mandi; it
complements life in this *mohalla* by giving a spiritual meaning to a
widespread sense of isolation and injustice.

Charity

It takes two men to lift the round, smoke-blackened metal cauldrons
called *deg* that line the courtyard. The cooks slide poles under a

small lip on the top so that they can lift the pots without burning themselves. They stagger with the weight, their teeth clenched. The pots are balanced on stones and charcoal fires lit underneath. Once the contents begin to bubble, an assortment of plates is piled on top of the *deg* to form a lid. Red coals are heaped around the rim and base, and they are left to cook unattended for an hour or more until the mixtures are stirred with giant metal paddles and spooned out to the hungry. Throughout the first ten days of Muharram there are dozens of pots sitting among the coals on the streets of Heera Mandi.

A basic type of food is given to the poor: *deg* full of rice, chick peas, and perhaps the odd piece of gristly mutton. Better-quality meals are given to other families in the area. The food is often excellent: the standard of the meat, the freshness of the vegetables, the fragrance of the rice, and the lavishness with which the *ghee* is ladled in are all an indication of status. Our next-door neighbor is distributing rice, peas, and pulses to the poor tonight. Three *deg*, one containing meat, another the best rice, and a third a delicious, thick bean soup, have been prepared for the community. Her children will take this food on plates to other local families. And they, in turn, will offer food another day. Their honor depends on not being seen as mean.

Those who can afford to, hire cooks who arrive with their *deg* and big bundles of wood and sacks of charcoal and set about cooking in the streets. The cooks are from the Nai caste—barbers who double as caterers—although the best do nothing but cook throughout the year. They provide the food for weddings and anniversaries, and they do a good business in Heera Mandi during Muharram. The highest caliber are in great demand because people with a reputation to foster do not want to be known for serving up substandard fare. Sharif Nai is the best cook in Heera Mandi, and absolutely everyone wants to get a taste of what he's cooking.

When Heera Mandi thrived and the customers were rich and plentiful, there was intense competition over feasts during Muharram and Eid. During Eid-al-Adha it's customary for a family to slaughter a fattened goat. In the past—even ten or fifteen years ago—more than one might be killed. The gossip would circulate that someone was slaughtering two goats, and in an effort to outdo its rivals, a household would counter by killing three.

In her days of earning *lakhs* of rupees Maha fed the poor and the entire area with a regal munificence. It is still remembered. How things have changed. Maha can barely afford to feed her children now that she is 35 and fat and her husband doesn't call. This gives many women satisfaction: they talk about it in lowered voices, professing a concern I don't think they feel.

Shamsa's New Family

Shamsa dances down the main road to meet me with laughter, shouts, and great breath-stealing hugs. She moved and has a new family: new aunts, uncles, cousins, nephews, and nieces. This girl has lots and lots of relatives whom she barely knows. She takes me to a room at the top of the house, a kind of lean-to on the roof. It's padlocked and she's obviously proud of it, swinging open the doors with a flourish. She looks at my face intently, and I don't have to feign surprise. The room is arranged very carefully. There's a sofa, a chair, a gray metal wardrobe with big gold hinges; the *charpoy* has a foam mattress and is covered in a cosy, fluffy green blanket. A shelf is smothered in red plastic flowers. It's not really Shamsa's own room: it's one she rents from the landlord by the hour for her business. She sleeps downstairs on a mat in a room with other women.

Shamsa's behavior is still very unusual. She puts on some loud music, drags a fan to the side of the bed, and switches it to maximum. Then she goes out of the room to speak to a customer. She returns a few moments later and asks me to look after two hundred rupees ($3). A thin, middle-aged man comes in beaming and asks me if I want hashish or something better. Maybe the "something better" is what keeps Shamsa in such a strange but happy mood.

When Shamsa returns she tells me that she came to Lahore five years ago with her family but that she no longer knows where they are. I tell her about my children and she grows quiet. "I have a baby," she says.

"A boy or girl?"

"A girl."

"Where is she?" I ask.

"She is in the sky. She died. I don't know why."

A few seconds later Shamsa is back at full volume, pulling me off the bed and telling me that we are going for a walk—another embarrassing parade though Heera Mandi, with Shamsa chatting to men and looking them directly in the eye. Her head is bare, her *dupatta* slung jauntily over her shoulder, and I'm intensely uncomfortable. Her behavior is particularly insulting during Muharram. For the next few years I'll have to walk around these streets, and I can't afford to have my reputation compromised in this manner. I like Shamsa—she's refreshingly different and a loose cannon in a very rigid social context—but I'm not sure if she's actively challenging an unfair society and making a stand for women or if she's just utterly mad. Whatever the case, being associated with such a wild young woman puts me in danger.

"All the men are looking," I say.

"Men are like that," she replies.

"Please wear your *dupatta*," I plead.

She won't.

Majalis

Maha's immensely fat cousin is hosting an evening *majlis* to which select local women have been invited. The cousin has a house near the sweepers' church; she's set up a *sabil* in the street that looks like a brightly lit little stage two feet off the ground, with colorful curtains and a canopy. Forty or fifty women are sitting cross-legged on the floor in her downstairs room. They are dressed in black, singing religious songs, and looking disconsolate. A few are crying. It's like a funeral. A group of boys are irreverently playing tag in the street, but the women surrounding me haven't noticed them, or the fight between two stubborn rickshaw drivers whose vehicles are locked together in what remains of the road. The women comfort one another as they sing. Every now and again someone glances up for inspiration at the religious pictures stuck all over the wall: highly stylized depictions of the body-strewn battlefield at Karbala; drawings of Mullah Ali looking strikingly handsome, with a full black beard and smoky eyes; a big colorful poster of Abraham and photographs of the tombs of Sufi saints. The food is served; the singing stops but there isn't the faintest trace of a smile.

The local community shows its religiosity and its wealth by hosting these *majalis* and performing the charity mandated by Islam. The better-off also organize parades, hiring a horse and all its trappings, and commissioning a *tazia*. Some of the participants in the rituals wish to pay their respects to the parade's patron, but many are simply desperate for food. There are more parades, more *matam*, more food distributions, and more *majalis* in Heera Mandi than anywhere else in the Walled City. In part this is because there are far more Shias here than anywhere else in Lahore, but it might also be that the people of Heera Mandi feel a greater need to prove their religious and social worth. They are guaranteed to put on an impressive show.

Another Muharram parade is passing through Heera Mandi tonight. Tarranum Chowk is packed and the road to Bhati Gate is lined with women as the procession advances in an energetic rush of men and horse. *Alam*, replicas of the battle standards of Karbala, take the lead. A great metal *panja* draped with bright shawls follows, then a smaller *panja* on a long pole wound with tinsel, flowers, and silky scarves. A white horse snorts as it passes by covered in finery, and following it is a *tazia* festooned in flowers. The women around me gasp and wail and some, overcome with emotion, begin to weep at the sight of the *tazia*. I ask someone what it is and they cry, "The cradle. The cradle. The cradle for the baby." An old lady takes my arm. "They killed him. They left him for seven days without food and milk. He was crying and they wouldn't give him water. And then they killed him. They fired an arrow into his little throat." She holds her head back and puts three fingers into her mouth to illustrate the method. The women around us nod. It's as if they have seen it happen themselves to a baby they loved dearly. This may be a confused memory of the martyrs of Karbala and the tortures inflicted on the family of the Prophet, but it is horribly real to the grieving women.

Hundreds of men are singing from song sheets outside the cinema, sad songs about martyrdom, death, and the killing of innocents and the just. Now I'm weeping too.

Zuljenah

Everyone from Heera Mandi is taking part in the *ziyarat*, the seeing and doing of holy things. To be accepted in Heera Mandi I also have to participate in the memory of Karbala. For the first ten days of the month we must watch the men beating themselves, we must touch models of tombs and cribs, and we must admire a colorfully dressed white horse as if it is divine. The horse represents Zuljenah, the horse of Hussain, which returned riderless to the Shia camp: a witness to the bloody martyrdom. This animal can reduce adult men to tears and women to a state of hysteria.

Maha is distributing refreshments to the poor lining up in an orderly queue. She bought a large drum of milk that's been sweetened and flavored with cardamom and nutmeg: one of her male relatives is overseeing the pouring of the milk while she looks on dressed in her best black chador. Along the main road a hundred or more men are performing *matam* before a beautifully adorned Zuljenah. Their bare chests are a bright shade of pink and some of the more vigorous have pounded away the top layer of skin so that their raw flesh is glistening in the sun, the blood running in rivulets over their stomachs.

The horse is well trained. It snorts with rolling eyes and stamps its restless hooves as the believers place their hands on his jeweled coat and kiss the flowers hanging over his back. I'm obliged to kiss his flanks, and the old man holding the bridle smiles and exclaims, "You're a Shia!" He hands me a fistful of bruised, slimy petals that have been made holy by their proximity to the horse's muzzle. My companions watch in satisfaction as I eat the blessed flowers. Happily, they do not taste of horse.

Ashura

The old city is quiet on the eve of Ashura, the climax of the Muharram commemorations. Most shops are closed, their metal grills pulled down and their wooden shutters padlocked. Rickshaws, vans, and motorbikes no longer clog the *galis*. Police barricades cordon off the main bazaars and *chowks*, and great tumbles of rusty barbed wire wind through the main streets. It's calm and peaceful, and boys have transformed the normally congested city into dozens of cricket pitches.

Ariba is accompanying me on my walk. She saunters barefoot by my side. I wonder if her lack of shoes is a kind of Muharram ritual of suffering and sacrifice. She's never struck me as a religious girl, so I ask her. She tells me that her shoes broke a few weeks ago, but that her mother says she doesn't have the money to buy new ones. Maha, however, has had enough money to buy me a new black cotton suit like her own especially for Muharram. I wince at the girl's broad, rough feet on the road and promise to take her to Babar Market for some new shoes as soon as Ashura is over.

We stop at one of the few open shops where an audience has assembled to watch sheep being butchered. Ariba's interest wavers when she spots a man selling *gola*—ices—from a cart on the opposite side of the road. *Golas* are a favorite with the children now that the hot season is here. The *gola walla* shaves pieces off a block of ice with a plane set into a wooden frame, packs the ice into a mold, and pushes in a sharp stick as a handle. The lump of reconstituted ice is pulled out of the mold and drenched in the syrup of your choice — something bright and wholly artificial — and the top of the fluorescent *gola* is dipped into sweetened yogurt.

A loud slurping accompanies our progress around the city. Ariba is relaxed and on familiar territory. She's not fussing with her *dupatta*. She walks with a streetwise confidence. She knows the nooks and the crannies of the lanes, taking me into the gloom of narrow alleys that snake through the city. At their base they are four or five feet wide, just enough room for two people to pass if they press against the walls. But above us, three or four storeys high, the buildings twist in, one upon the other, so that in places there's only a three-foot gap between the houses and the sun never reaches beyond the windows of the topmost storey.

We meet one of Maha's many relatives near the bazaar, and he accompanies us through Tibbi Gali. For once the place is closed. Most of the doors are locked and only a couple of women are discreetly looking for business. Nazia's madam sits on her doorstep, but Nazia isn't here. She has gone away and will be back after Ashura. As we emerge into the shuttered shoe market, Maha's relative tuts and tells me it's a dangerous place. "It's too dirty," he says. "The women there cost only two hundred rupees." I suppose it would have been less dirty if they, like his female relatives, had charged two thousand ($34).

In a side street, Malika, Tasneem's landlady, walks toward us with a group of *khusras*. All are dressed in silky black *shalwaar kameez*. They look like a flock of big, shiny crows flapping up the lane. Malika caws and gives me an intense beady look. She asks when I am going to visit. Ariba stares at her in amazement because Malika has seriously overdone the scarlet lipstick and diamanté, and even Ariba, with her great fondness for glitz, knows that the look isn't in keeping with Muharram. I'm glad when they rustle and squawk off home.

The biggest procession starts on the evening of the ninth of Muharram and involves the entire inner city, skirting Heera Mandi as it winds its way slowly along the streets and *galis*, stopping for Zuljenah to visit mosques and mausoleums and the houses of the better-off. It will continue throughout the night and finish late on the tenth, on Ashura.

Masses of people crowd into the part of the old city where the procession begins. Most of them are men jostling in the dark and pushing through bottlenecks in the *galis*. Maha, Nena, and I squeeze our way into a mausoleum where two hundred women and a few frail old men are waiting in comparative safety. Everyone is performing *matam:* beating their chests and chanting the name of Hussain. A shout from the entrance tells us Zuljenah is coming, and there's an immediate wail and a dramatic surge in the chanting and the force of the beating. It's difficult to believe that the horse can enter this packed room, but it does. Zuljenah jumps and bucks through the door as the women cry, fighting to get close to it. Maha elbows her way into the mob and I lose sight of her. Nena stays close to me, but she's not aware of me by her side: she's looking in adoration at Zuljenah, tears coursing down her face, her breathing fast and shallow. Dozens of women hang onto the horse; others swirling around it until Zuljenah is maneuvered out of the mausoleum and the women are left dazed and weeping.

A wild-eyed Maha returns. Her neck is red and she's bareheaded. Someone had taken advantage of the chaos to try to tear off her gold necklace, but the thief had only managed to snatch her best *chador*. Maha fumes, gingerly touching her neck where the chain has cut into her skin. An old lady clucks in sympathy and gives Maha a shawl to wrap around her head so that she won't be shamed by appearing unveiled in the street. Maha undoes her necklace and stuffs it into the great vastness of her bra. No one will ever be able to find it in there.

We follow in the wake of the white horse, the crowds around us working themselves into a frenzy. Long lines of men push through the middle of the spectators holding knives and chains above their heads. Those going in the direction of Zuljenah are carrying clean blades pointed upward—they look brand-new and a few are still wrapped in newspaper. In the opposite direction come the men who have used their knives. Some walk, many stagger, their chains and blades held even higher but now covered in blood that runs down over their hands. They're stripped to the waist, their backs a mess of wounds and blood soaking their white *shalwaar*. A few are so badly injured that the blood-sodden fabric clings to the back of their thighs. It's a grotesque but triumphant and profoundly moving procession of men bearing their symbolic trophies of battle and martyrdom.

In the path of Zuljenah men hold the wooden handles of their chains and swing them around, first one way and then another, so that blades cut into the flesh of their backs. Fine droplets of blood spray the crowd, and our hands and faces are speckled with dried blood long before dawn.

A loud, deep thud of *hath ka matam* is coming from a large mosque filled with two or three hundred women all crying for Hussain. An announcement comes over a loudspeaker that a child has fallen under the horse's hooves and been trampled to death. The women beat harder and call more desperately for Hussain: another child has been martyred.

It takes a full hour for Zuljenah to arrive and the *matam* is becoming painful. I don't know how long I'll be able to continue. The horse arrives and is led immediately into a side room; the tired Zuljenah is being replaced because it's impossible for one horse to parade nonstop around the city for twenty-four hours. Dawn is spreading through the open roof before the fresh, enormous new Zuljenah bursts into the mosque, and by the time it has pranced out, two dozen woman are lying unconscious on the floor.

The finery is being removed from the retiring Zuljenah. Maha drags me into the room despite my protests that it's dangerous to be in such a small, overcrowded place with such a large, exhausted, and nervous horse. The animal is swamped with adoring people kissing its flanks and rubbing it ecstatically. A large pink *chador* is taken off the horse's back, torn into little strips, and distributed to waiting women. These holy and blessed strips are tied around their wrists to bring them luck. As we walk away Maha delightedly presses a strip to her lips. I hold my own drenched piece at a distance.

"What's the matter?" Maha asks.

"I don't like horse sweat," I reply with a laugh and, in an instant, she turns on me.

"Don't say that. It's such a bad insult," she berates, deeply offended. "It's not funny. It's not horse sweat. It's beautiful. It's *ziyarat* perfume."

B ack in Heera Mandi I sit in the bazaar with Iqbal, enjoying the morning sunshine and the release from the crowds. His wrists and forearms are covered in deep gashes, as if he has made a ham-handed attempt at suicide. He explains that it's the responsibility of family members and friends to ensure that *matam* never becomes life-threatening: it's hard for a man, mindful of his honor and his masculinity, to stop the flagellation voluntarily. Others have to intervene. One of Iqbal's friends was particularly enthusiastic and difficult to restrain during his *matam*, and Iqbal is bearing deep scars as proof.

Dozens of youths walk by in their blood-soaked *shalwaar kameez*. They stink of stale sweat and congealed blood but refuse to change. They wear their bloody clothes as marks of honor, swaggering with pride; you can see them wince as the material touches the wounds, but they pretend it doesn't hurt. We walk with some of them up toward Tarranum Chowk. The area is closed off and police are everywhere. The road to Bhati Gate is a seething mass of bodies. From the Tibbi police station we can hear the chants and see the flying chains as the procession continues through the old city.

T he procession will end in Karbala Gamay Shah, the most important Shia shrine in Lahore. We leave after dark on Ashura and, in the gloom of a Heera Mandi *gali*, one of Maha's many cousins approaches us with news. He saw the *dudh walla*—the milkman—give Ariba thirty rupees (thirty-one cents) this morning. Maha begins to shout at Ariba using the foulest language. "You are a dirty *gashti*. A *gandi kanjri*."

Ariba cowers and stumbles up the lane without saying a word in her own defense. Maha is fuming at the terrible shame of it.

"Why is the *dudh walla* giving her thirty rupees?" she screams. "He wants to touch her breasts." She lands Ariba a fearful thump on the back of her head that sends the girl flying to the ground. "She's out of control. What can I do? What can I do when my daughter is a dirty prostitute?"

At eight o'clock the streets around Karbala Gamay Shah are thickly lined with men. A large section of grass has been set aside for the women, and the proceedings are patrolled by more men in blood-stained *shalwaar kameez*. On the tenth of Muharram dried blood on men's clothes gives them a kind of official status. Women and young children sit on the grass gossiping. I recognize several faces from Heera Mandi and stop to speak to them. In their conversations with other women, they never mention that they come from Heera Mandi: they say their homes are in Karim Park or some other respectable suburb and I am their friend from America. A friend from England, it seems, does not have the same cachet.

Maha draws her *dupatta* tightly around her head and mumbles, "Adnan's sister is over there and so is his mother." A couple of fat, grim-looking women are staring at us. Mutazar and Sofiya play around Maha as she smiles false smiles and repeats, over and over again, that she won't be shamed. She's not going to lose honor by being intimidated by Adnan's official family. The children's grandma and aunt neither wave to them nor smile, nor take anything but hostile interest in them and their mother.

The procession is running late and an announcement informs us that it will arrive in half an hour—which, in real time, means that we could be under the cold eye of Adnan's family and sitting on this hard earth with its sparse grass until midnight. It's too much of an ordeal, so we dust ourselves off with all the dignity and style we can muster and we catch a *tanga* home. Maha is silent and only her shaking lets me know she's crying. She's rubbing the strip of fabric from Zuljenah's *chador*. We're traveling along Mony Road, and as we pass Adnan's houses she sobs aloud.

Training for the Bazaar

The hot season is gathering pace and the temperature is rising by the hour. Those with air coolers have switched them on. The white flowers cloaking the climbing plant beneath my window are wilting and edged with brown. The stone floor of the mosque has absorbed the intensity of the sun's rays, and a path of sacking has been laid and soaked with water to cool the worshippers' feet. Those who stray from the path must have developed impenetrable calluses and thickened soles; the others hop.

Mushtaq, the drug pusher and champion pimp, is preening in the courtyard. I can see him from my window. He saunters around in tracksuit pants and a tight T-shirt. He's grown his beard and looks splendidly dark and moody, but he has a disconcerting habit of massaging his genitals in an absentminded kind of way, as if they are too weighty and uncomfortable for him to manage without support and frequent rearrangements. Either that or he has crabs. This could be a problem for the girls working in the thin pimp's brothel because Mushtaq likes to try them all—often.

Nena is training for the bazaar. By the standards of the *mohalla's* classical tradition, this training is starting six years too late, but an intensive short course in dancing is all that's required in today's Heera Mandi. A dance master visits the house every day; this evening they are working on a modern routine—Bollywood meets Western cabaret. I don't like it, but it's very energetic.

The dance master is young. He lives outside the *mohalla* and is wearing Western clothes: a black shirt and tight gray trousers. All the dance practice has given him a very pert bottom, and Maha is in hysterics whenever he swings his hips to show Nena just how the movement should be executed. Maha doesn't let the teacher see her amusement. To his face she's very respectful, calling him "Master *Jee*" and "Master *Sahib*." She saves the laughs, the grimaces, and the hilarity for his bottom.

After a few weeks, classical training has replaced the Hindi pop routine, and Nena is progressing really well. She wears her heavy *ghungaroo* and executes some of the basic movements with a supple and lovely elegance. After almost two hours of practice she collapses on to the bed, flushed, her feet aching because of the constant stamping. She rubs her soles and her instep and winces.

Employing a dance master and musicians is an expensive investment. Maha can't afford it, but it's important for Nena's future—and for that of the whole family. Maha is trying to reestablish her prestige in the community by becoming a patron. She can only sustain this image with cash, and for the last month, the musicians and Master Jee have not been paid.

The men assemble on the floor while Maha sits on the mattress and arranges her *dupatta*. The musicians refuse to join me on the sofa because it would not be polite, and besides they're asking for their money, which means that they're required to be deferential. Maha is embarrassed and tries to cover her discomfort by laughing with me and joking with the men. The social pressure to pay is intense. They know her. They're part of the same community. They're not asking directly for their money, but the message is clear: if she wants to hold an honored position in Heera Mandi she will have to pay, and do so quickly. The fee is not much—fifteen hundred rupees ($25) for the month, to be shared by Master Jee and the musicians—but it's money that Maha hasn't got. She promises to pay them soon—by tomorrow—pleading husband problems, and they leave. I don't know how long she will be able to keep up this pretense and this show of status when she can barely afford to pay for the electricity and water supply to her crumbling home.

Honor

Life in a brothel is vicious. It's not just the clients and pimps who are vicious, but the other women, too. There's very little solidarity here, and the women rarely acknowledge that they have any common interest apart from Shiism. I've never heard a woman in Heera Mandi compliment another woman outside her own family without also following this with deeply derogatory comments. Others are ridiculed as old, ugly, fat, cheap, and lacking in sexual allure, artistic style, or refined manners. To their faces the women are sweet and charming, but as soon as their supposed friends leave, they are trashed. I wonder what they say about me. I think I can guess: I'm getting old, I'm looking thin, I have no jewelry, so I must be of no value. To prove the point, I don't have a man.

Maha hates me visiting anybody else, and I have to be careful about the division of my time. According to her I can have only one friend. My loyalty must lie with her completely: friendship is available in limited quantity and can't be shared. Last night I had dinner at the house of an elderly local woman. She retired from the business years ago but still lives in some style. She was a great *tawaif*—a classical dancer who trained from the age of 7—and she saved enough of her money to live a comfortable life in retirement. We had a lovely meal and looked at old photos of her . . . and I have just made the mistake of mentioning the visit to Maha.

"What?" she shrieks. "You went to that Pathan's house? That old fat one?" She huffs and looks indignant. "What did you eat?

"*Rotiya* and *kofte.*"

Maha is aghast. "And have you got a pain in your stomach now?"

She prods me sharply. "She does black magic. She's probably poisoned you. Never eat there again." Abandoning her vow to stop the drugs, she unscrews a bottle of Corex and drinks it in a couple of giant gulps.

A vile smell wafts from the toilet in the corner of the room and Nisha is ordered to mask the odor with air freshner. Maha displaces her anger from me to the smell and then to the woman in the rooms above. "It's that *gandi* Pathan upstairs. That smell is from her dirty bum. She's never clean. She has a filthy *kusi*. She's a cheap, *gandi kanjri.*"

It's true that the woman upstairs is a prostitute, but how Maha can abuse her on this count must take some sophisticated intellectual gymnastics. Everyone in Heera Mandi maligns everyone else as a prostitute but when they themselves are called the same names, they cry and rage with the shame of it: all the neighbors have heard them being called a *kanjri* or a *taxi.* As if no one knows what happens in Heera Mandi, as if they're all respectable housewives. Most have never been able to break out of a culture built upon male privilege; even when they can see, in the unremitting unfairness of their lives and those of their mothers and grandmothers, how those ideas consign them to a social ghetto of stigma and shame. Ironically they damn other women caught in the same desperate situation with the very words and concepts that will be used by society to damn themselves.

It's a constant effort to keep up appearances, maintain status and honor, counteract insults, and fight enemies and detractors. The most vicious and loudest arguments are saved for neighbors: for anyone within shouting distance. Two families in the courtyard have been locked in bitter conflict for several days. They scream insults at each other, "Tell your mother she's fat," "You're sick. You're going to die." "You are so ugly."

Both households tell me about the stupidity of the other. "They're jealous," they say. "Look how ugly they are. They can't dance. They're cheap prostitutes. They're the children of a dog." The fights are intense and fun and exciting: they punctuate a day of waiting and eating.

In Heera Mandi's complex social code women with *izzat*—honor—buy real jewelry and have extremely large fridges, televisions, video players, tape decks, and mobile phones. They donate money to the poor and needy in the community, pay for feasts during religious rituals, and are the patrons of musicians and dance masters. This is what all Maha's hard-earned money was spent on. A woman who can display and redistribute a large amount of wealth within the community will have high *izzat*—providing that she also talks of her clients as husbands, keeps her prices very high, scrupulously observes the Shia religious calendar, and keeps *purdah* and veils when appropriate. If she can achieve all this she will win the respect—and the bitter envy—of the community.

On first acquaintance all the women claim that they are expensive, and they inflate their prices to astronomic figures. To admit they sell sex for five hundred rupees ($8) would lower their *izzat* and so reduce their real price. This emphasis on creating an invulnerable and successful image means that women perceive everyone else to be enjoying good business and having a happy life. Everyone else's clients are thought to be rich and generous and other women's husbands treat them with love and respect. Women think it's only their own business that is bad, it's only they who cannot afford to pay the rent, it's only their own husbands who are cruel. They don't believe me when I tell them that the lives of other women are equally blighted, and they say that I mustn't let anyone know of their own difficulties. They are "ten-thousand-rupee women" ($169) and they have a reputation to preserve.

Maha is upset, ranting about insults and dissolving into tears. Adnan hasn't called for a week, so this afternoon, she went to his house in Mony Road to ask for money to pay the rent and feed the children. One of the servants took the message to him while she waited outside in the rickshaw. She's telling me the tale in strangled sentences.

"I wasn't like a human being. I was like an animal. I sat in the rickshaw in the sun for more than an hour and Adnan wouldn't come. He knew I was there, but he kept me waiting like a dog begging under his window. My *izzat* is nothing. He gave me five hundred rupees and said 'This is for a week.'" She weeps with rage and humiliation. "I have the *izzat* of a dog."

Hasan

A boy earns a few rupees by running errands in the courtyard, dashing between the houses and the local tea shops with trays of tea and food. He skids to a halt by my side whenever he sees me, and I often ask him to fetch bottles of mineral water. He's an enthusiastic worker, sometimes too enthusiastic and too overwhelmingly earnest, asking, "Do you want me to open the bottle? Is it cold enough? Is it nice? Do you want to eat something? Do you want a Pepsi? I know, you want a bottle of 7-Up. Tea? No? I know, you want coffee." At times, I just want him to go away. He's very short with a thin face, worried brown eyes, and close-cropped black hair. He says he's 15 but his big new teeth and squeaky voice give his age away: he's about 10 and his name is Hasan.

Like all the children of Heera Mandi, Hasan has a difficult family background. His mother is a low-class prostitute, and he's one of the four sons she can't afford to raise. Hasan's elder brother has had a little schooling, but Hasan has had none: he's illiterate and will probably remain this way for the rest of his life. Where government schools are available, primary education is technically free, but in reality, the poor find the cost prohibitive. They have to buy books, pencils, uniforms, and transport. Most importantly, they forfeit the money the child could have earned by working.

There would be little point sending Hasan to school even if his mother could find the money. State schools in this part of Lahore offer the poorest education, and, assuming he gained some qualifications, the prospect of a good job would still be slim. Few employers want to hire workers from the wrong side of the city walls, and universities and colleges hold the people of Heera Mandi in contempt, as if they were stupid as well as poor and unlucky. In Pakistan, perhaps more than in many other societies, success depends upon whom you know. The residents of Heera Mandi know a lot of people, but in a way that is profoundly harmful to their own reputations and prospects.

If brightness, intelligence, and hard work were determinants of success in this world, Hasan would grow into an important man. The tragedy is that, barring some miracle, he'll run errands all his life, sink into heroin, or, if he has aspirations, he'll follow the example of other ambitious men in the *mohalla* and become a drug pusher or a pimp. He's an enchanting child: he reminds me of my son and I've made a fuss over him. I've also made a naively stupid mistake: I told Hasan he was a lovely boy, I took an interest in his life, I hugged him when he was sad, and I praised him when he worked hard. This afternoon when he came to my room to see if I wanted yet more tea, I thanked him but said that I'd already had plenty. He looked around, examined a few things, and asked some questions. After a while I said he should leave because I had lots of work to do. As he departed he threw his arms around me and I patted his soft-bristled head and returned the hug. To my horror, my little friend had an erection that he pressed insistently into my thigh.

It is now early evening and I'm sitting on the roof terrace recording the day's events. As I write, Hasan is playing cricket below me in Fort Road with other small boys who aren't competent or old enough to play with the youths on the grass by the mosque. He's looked up countless times, searching for praise for his poorly developed batting prowess and waving once or twice when he thinks I'm glancing in his direction. I've ignored him: I don't know what else to do. Poor Hasan has no chance growing up in a world in which sex and money are the only currencies in which affection is expressed. This precocious but sweet boy has been schooled watching his mother with her customers and has learned all the most damaging lessons.

Ariba's New Shoes

Maha has overdosed on sleeping tablets. One day she'll miscalculate and she'll die. She took 10 Eighty-One tablets last night after her humbling by Adnan and then she washed them down with Corex. These are potent sleeping tablets, all too familiar to the women of Heera Mandi. This morning Maha's speech is so slurred that she can barely speak and she can't even sit up in bed.

By evening Maha's drowziness has worn off and she's mellow and happy. Her relaxed mood is catching and the children are laughing and lively—except for Ariba who is, as usual, behaving like a harried and reluctant servant. Maha starts complaining again that Ariba is acquiring a reputation for hanging around the streets. It's an unfair thing to say because it's always Ariba who is sent out alone to buy food for the family. Tonight I'm going with her to buy *paan* for her mother and *rotiya* for the meal. In the shop by the cinema the *paan walla* gives her a free special *paan* and she tucks it into her clothes with glee. Then we watch the *roti* being made. A young boy helps his father by rolling the dough into little balls and patting them into circles with his fingers. There's a long line in front of us, and when we return home Ariba is scolded for taking so long and getting the wrong type of bread. "We wanted big ones, not little ones," her mother complains and tells her to go out again for more.

Ariba sits and sulks.

"Go on, bitch," Maha shouts.

Ariba mutters curses and her mother shouts louder. Ariba grabs her *dupatta,* crying, and slams the door behind her. I can hear her running down the steps. She's shouting, "You never loved me."

I watched Ariba for an hour this afternoon while she was in the courtyard, still barefoot. She was patiently looking after the small children: organizing games and races; hugging them when they fell over; adjusting their clothes; and dragging a wriggling and bellowing Mutazar away from the peddlers selling sweets and drinks.

She never receives any praise or acknowledgment for this. She's never thanked for running errands. Her work has an invisible quality.

Ariba shuffles around Babar Market in a pair of her mother's old, too-large shoes. She stands outside a shop and admires all the most inappropriate sandals. "I want some red ones," she tells the shopkeeper. "Pretty red ones. Ladies' shoes." I know the ones she means. They're in the window: red satin, encrusted with diamanté with four-inch stiletto heels and thin straps. She thrusts her broad, cracked feet into them and tries to walk. I suggest something more ordinary, and eventually we settle on a pair of white sandals with a low heel. They're not a suitable color for Heera Mandi's streets but Ariba loves them.

We spend the rest of the afternoon shopping and eating snacks. I've told Ariba she has one hundred rupees (not quite $2) to spend as she likes. Her choice of purchases and her shopping priorities say a lot about her. She has bought a cheap gold-colored necklace for herself and then she spends a long time in a drugstore deciding which talcum powder to buy. She smells every one three or four times, and decides, in the end, to plump for something floral because it's the cheapest. She's going to buy a present for Sofiya with the rest of her money. She's searched the entire market for just the right kind of hair clips. They are the palest blue and decorated with a tiny, pretty mouse in a pink dress. She's delighted with them. She looks at me with shining eyes and asks if her little sister will like them.

This evening Sofiya is wearing her hair clips and Ariba is wearing her necklace. You can barely see it's gold because Ariba has used so much of the talc that her neck is white and the necklace is dusted with powder. Her older sisters sit on the bed with me and one says, "What's that terrible smell?" They put their hands over their noses and mouths and turn to look at Ariba. "Isn't it horrible." Everyone laughs and Ariba scrambles from the mattress and runs away into the night.

Torturing Maha

Maha is in a foul mood. I've arrived in the middle of a nasty fight. A half-eaten lunch is lying on the floor and the children are hiding in the bedroom or in the corner of the kitchen. Maha is ranting. "They torture me. They'll kill me. I do everything for them and then they insult me."

"Louise," she says, "My head." She holds it in despair. And then, despite her size, she springs up, runs into the kitchen, and begins to beat Nena.

I don't know what Nena has done wrong. I suspect she's done nothing but be lippy in the manner of 13-year-old girls. She's sitting with her legs pressed tightly together and her arms rigid. She tries to give me a smile, but when I stroke her hair she begins to sob.

"I have to take those drugs," Maha pleads when she's calm enough to sit quietly on the mattress. And it's true: she has to take them to be able to cope. She's alone in these four rooms with her children all day, every day. She waits for a man to turn up and give her companionship and the money she needs to feed her children and pay the rent. These aren't things she can provide for herself without selling sex.

When she takes her medicine she relaxes because she doesn't have to think. She sleeps as Mutazar and Sofiya run wild and out of control; as Ariba, caked in filth, hangs about the streets; as Nisha quietly fades into the background; and as Nena cleans and watches her mother for signs of life or death. And then Maha returns to the world of consciousness and her unhappy family, her absent husband, and her lack of money and tries to reimpose control. She's insanely bad-tempered, the family is wracked with tension, and so she opens the Corex and counts out the sleeping pills. But it's never enough, never the answer, because when she wakes up nothing has changed.

I tell Maha not to take so much of the medicine. It's a useless bit of advice and a cruel one too because I can't provide any better form of relief. I tell her I'm leaving, and she asks me where I'm going.

I say I'm going shopping in Gulberg. It's where the rich Lahoris shop. She asks if they can come too and in a mean-spirited moment I say no, I have to go alone because I have other things to do. I feel incapable of dealing with all of Maha's problems as well as my own. Increasingly, I'm out of my depth and sometimes I have to walk away.

The Bangladeshi Family

I have some gruesome photographs taken on my last visit to give to the Bangladeshi family: two shots of the mother and father, an unflattering portrait of the mother that catches her bad teeth in a very harsh light, and a group photograph taken with Shamsa in her green glitter lipstick. They're thrilled with them and I sit on the *charpoy* with Bilquis, the daughter, to inspect the pictures. Five or six children were in the room when I visited in December but Bilquis is the only one here today. She's so fragile and pretty with her fine bones and wide eyes. She says she's 15 although she can't be more than 13.

"I was married two months ago," she says. "I wanted you to come to the wedding, but you weren't here."

She shows me a picture of her husband. He's in his twenties, with bristly hair and a downy moustache. I congratulate her and say he has a nice face. She says, "Yes. He's a good husband and he makes bedsheets."

The girl's mother is sitting on the floor playing a board game, ludo, with a man who is also from Bangladesh. The game is engrossing and a lot of low-value notes are changing hands.

"Did you do the *ziyarat?*" Bilquis questions as she shows me a strip of material wound around her wrist. It's a piece of Zuljenah's *chador.*

"Yes, many times. And I touched Zuljenah."

"Ohh . . . now your luck will be really good."

Unlike most families in Heera Mandi, this family is from the majority Sunni branch of Islam, but they still take part in the rituals of Muharram as if they are Shias.

Bilquis says that I must want to listen to some music, and she whisks a thick rag off a tape player. It has colored, rotating lights and I remark that it looks nice. The family agree. The player is of such significance to them that it is immediately re-covered with the rag to stop it being spoiled by the dust and we listen to the muffled music through the material.

The father has something to say to me and keeps trying to interrupt his daughter and to speak in English. A man comes in, stands by the *charpoy,* and offers me a thin joint of hashish. The family greet him as if they know him well. He is fortyish, bony, and some kind of low-skilled worker. Bilquis's mother gets up and goes with the man into the other room, which is in darkness, and closes the door behind them. I was wrong to think she was no longer in the business. The lonely ludo player enjoys his hashish and glares at me intermittently. Five minutes later the woman is back playing ludo, and the client leaves scratching his balls and adjusting the drawstring of his *shalwaar.*

"Where is Shamsa?" I ask, wanting to find out why she has left the house.

"She went to some other place," the father replies.

"Why?"

He shrugs and the daughter rolls her eyes. "She was crazy, *bad-tamiz.*"

"What is your business?" the father asks. I explain that I work in a university and that I'm writing a book about Heera Mandi. He knows this already and I wonder why he's asking me again.

"Do you have any other business? Business that gives you trouble with the police?"

I shake my head.

"You've been sleeping at Maha's at night, haven't you?" he asks.

I confirm that I have, occasionally. He's observant—the mark of a successful pimp. He digests this information and continues. "What do you think of this business in Heera Mandi?"

"It's the life here."

He chuckles and shifts around. "What's your price?"

The ludo has stopped and everyone is looking at me.

"Very high," his wife laughs.

"Come here," the husband beckons and goes into the other room. I hesitate but still follow him. It's dangerous and I'm scared. There's no easy escape route. He opens the shutters. There's nothing in the room except a dirty mattress on the floor under the window.

He talks to me in a quiet voice. "Do you do the business?"

"No. I'm married," I lie. "My husband is in Islamabad."

"If you're in the business, I can get you a very good price."

I thank him for his offer but say I'm not looking for clients. He smiles and, bizarrely, takes my hand and shakes it. As we leave the room, a new visitor walks in with the mother and the ludo is interrupted again.

I've been making discreet inquiries about the family, and the stories are consistent. The family trades in children—especially Bangladeshi children destined for prostitution or the marriage or domestic servant market. That is why the gang of children had been whittled to one: there was a problem with the police last month. "There was a little girl in that room," two women tell me. They point to the shutters of the room with the dirty mattress. "They sold her to four or five men. It was rape. The woman downstairs heard her crying for help." The story is that the police came but there was nobody there.

"What happened?" I ask.

"A bribe. Or they sold her quickly to some other place. I don't know."

The Kanjar don't like this family: they're outsiders—rough and villager-like. They're also suspected of truly scandalous practices. I've been told about these practices at long and horrible length: they catch pigeons on the rooftop, slit their throats, and make them into curry.

Operations

There's a strange atmosphere in White Flower's dark, ever-crowded room. The shutters are rarely open. Perhaps it's because it's too hot outside or perhaps it's because it's not worth opening them. There's nothing to see apart from the gray building four feet away on the other side of the alley.

White Flower is lying on the bed having a massage from an elderly *∂ubanna walla,* a masseur, who is working deeply and vigorously on her legs. White Flower's eyes are closed and there's a trace of a smile on her lips.

New faces fill the room. I recognize only White Flower and one other younger *khuʃra* called Samina, who is always friendly. I'm given a seat and White Flower stirs from the massage. She gives the quickly departing *∂ubanna walla* twenty rupees (thirty-four cents) and shrieks a greeting. She undoes her long hair, flicks it around, apologizes because it's covered in oil, and asks me to comment on how it has grown.

"Have you seen Tasneem?" I ask.

"She's gone," the others reply, all with slightly different versions of the story.

"She's gone back to her village."

"She grew a moustache," White Flower explains.

"And then she got married," someone adds.

"Married?" I query.

"To a man. It's a boy-boy marriage. Man-man." White Flower half-screams.

"She had lovely hair," Samina sighs.

Two young clients are sitting in the corner and White Flower introduces them as "guests." They're clearly intimidated by being here. They say nothing and stare at the floor. Perhaps they will cheer up after I've gone, but I guess that won't be for some time because I'm presented with food and a bottle of Coke.

Samina sits on my left plucking at her stubble with an enormous pair of gold tweezers. The seat on my right is taken, in turn, by different *khusras* who want to have a look at me and quiz me about beauty treatments. Is there a medicine in London that can get rid of facial and chest hair? Can you get a cream to make breasts grow? How can you get very, very long nails? Is there a wonder solution you can paint on like nail varnish? How can you get really white skin? I don't know why they're asking me all these questions. I'm the least well groomed person in the room.

The two prettiest *khusras* are made to stand in front of me, and I'm asked to judge who is the loveliest. One of them is exceptionally attractive. She looks like a woman. She has high cheekbones, soft features, and a slim figure. She says she wants to perform a dance for me providing that I do one first. So I take off my *dupatta*, make my usual enthusiastic attempt, and then leave it to the professionals. Except that there's a little delay because the music has to be switched off while the *azan* calls the men to prayer. I sit down in the quiet, and for the first and only time the *tamash been* speak. In the middle of a *khusra* brothel these young clients tell me to show respect and cover my head. I joke that no one else is doing so, and the *khusras* laugh and shrug and says it's not necessary: they're only half women and so are excused from women's obligations.

The beautiful *khusra* dances wonderfully well and the pious *tamash been* have lifted their eyes from the floor. When she has finished I pay her a very sincere compliment: I say she's lovely—just like a woman. Yes, everyone agrees, she's just like a woman.

"She had an operation," Samina adds with emphasis. It happened four years ago. It was in Karachi, in a hospital, and she paid the doctor fifty thousand rupees ($843).

"After the operation a man becomes just like a woman," Samina states. "They don't have a penis or anything else."

"Just like a woman," I repeat. "But they don't have a *kusi*." It's the wrong thing to say. For a few seconds there's silence. The *khusras* look away from me and then there are some giggles.

"Who needs one of those?" someone quips.

Cinema

Didi is a young and unusual *khusra* who loves to dye her hair. Today she's a startling, golden blonde. We've started meeting occasionally in a tea shop on the main road. It shows Punjabi films to an exclusively male audience on a little television at the back of the shop. There are lots of these tea shops cum video rooms in Heera Mandi. It's a good way of pulling in customers. Punjabi films are popular because the customers in these places are largely poor laboring men. Punjabi is their first language—the language they are comfortable with and the language they use when they relax. A language hierarchy divides Pakistanis: lower classes speak regional tongues like Punjabi or Pashto, and the middle and upper classes speak the national language, Urdu; the cleaner the grammar, the higher the status. The richest, Westernized people speak English, sometimes with a British accent, but more often these days, with an American one, proof of an expensive foreign education.

Heera Mandi has another type of informal cinema too. At least one place in the *mohalla* offers pornographic films. It's just a little room with a shuttered front and no windows. With no air-conditioning or fan, the audience bakes in summer, emerging after the show staggering, dazed, and dripping with sweat. My guess is that they don't even notice the temperature.

This morning Didi is sitting in the tea shop dressed in men's clothes and a lot of fake gold jewelry. Her newly yellow hair is scraped under a green baseball cap that's slightly too small. She must have overdone the bleach because the hair at the nape of her neck looks like a nest of thin, crisply fried egg noodles. The film on show is a 1980s Punjabi production. The star is a fat, middle-aged man with dyed black hair who runs about making a lot of puffing noises.

"This is rubbish," Didi comments. "I like Hindi films." The rest of the audience must not agree because they are sitting, silently engrossed, while their sweet tea develops a thick brown skin and grows cold in their cups.

A couple of hours later we are sitting on the steps of the Tarranum Cinema. Didi is eating *paan* and there's a bright red semicircle of *pik* stains radiating out on the ground in front of her. She's working with a young cucumber seller who has stationed his stall by the side of the road, about twenty feet away. The cucumber seller is encouraging his customers to look toward the steps. Didi smiles at groups of passing boys and youths. Some of them respond by coming back for a second look. Two or three youths have stopped and there's a vulgar exchange between the boys, Didi, and the cucumber *walla.* He's waving a peeled cucumber around and everyone is laughing. The boys meander off toward the shoe shops and the cucumber seller is angry. He's pointing at me and telling Didi that I must go. "They thought you were with her," he shouts indignantly.

Salma

Tibbi Gali is transformed now that the saddest days of Muharram have ended. By five in the afternoon the heat has gone out of the day. The doors that were locked last week now open into dark little rooms and heavily made-up women are sitting on the doorsteps. I've been trailed by Ariba and Mutazar, and the women call and ask me who I've brought. Are they my children? I say that they're my sister's children—Maha's children—and that they live by the Badshahi Masjid.

Nazia is sitting on her doorstep. She's just the same in her old blue outfit but she's wearing a different shade of lipstick. She grins at me and the color smears over her teeth. A little farther down the *gali* an elderly woman beckons. She must be about 60. She has dyed hair, heavy makeup, and thick glasses. In most large brothels in any other Asian country she wouldn't be selling sex because there wouldn't be any demand for her. I used to think that the old women in Tibbi Gali were madams, but I've had to revise that opinion. Some are, but there are many other desperate women who are still working into their fifties and sixties. This woman speaks to me in Punjabi. I talk back in Urdu, but her own responses in Urdu tail off into Punjabi and it's hard for me to follow what she's saying. I know she came here as a child, that she was sold into Tibbi Gali and had no choice. This must have been fifty years ago. I pause and try to imagine what it must be like to have been traded in Heera Mandi for fifty years — for all your adult life and more. I can't possibly imagine it. And yet, in the shade of her tiny room, with its *charpoy* barely hidden behind a curtain, this woman, who is the same age as my mother, can still laugh and tease me because I don't have a man.

Salma is another older tenant: she has a room at the lower end of Tibbi Gali where it opens onto the road to Bhati Gate. She must have to watch carefully or pay very large bribes because she is so near the police station. Salma is relaxing on the floor, her head resting on a pillow that long ago became flattened with use. It's hot and there's no fan, but at least it's cooler in the dark interior of the room. The *gali* is narrow here and snakes up and around a corner so that the sunlight never enters the rooms. Her working quarters are similar to those of poor prostitutes in the big brothels of India and Bangladesh. I've been in similar places in Calcutta, Mumbai, and Dhaka — the difference is that Salma is a lot older than most of the women working there. Her room has no electricity and no running water. She waits for customers on a blanket on the floor, and behind a threadbare curtain there's an old mattress that she uses for the business.

"Sit down," she says, smoothing the blanket.

She is, perhaps, 45. It's hard to tell. She claims to be 35, but everyone lies about their age. She has four children and her home is a long way from Heera Mandi: she comes here every day and she says it takes over half an hour in the rickshaw.

I ask her when she started in the business.

"Ten years ago. My husband left and I had no money and no one to help me. I was forced to come here." She thinks for a moment. "Can you help me? Can I get a job in your country? I'll work hard.

I can work in a factory sewing clothes or I can do cleaning. I hate this work. And business is bad."

I tell her about the problems of visas and work permits, but she persists. She only drops the subject when a customer stops to haggle over her price. The *chae walla* gawps at me as he hands us cups of syrupy tea and Salma explains why I'm here: I'm her sister, from America, and I'm writing a book.

"Can you read and write?" I ask her.

She laughs and shakes her head. "What a useless thing that is in this place."

Salma is a popular woman. She's not popular in a business sense—she doesn't have very many customers—but lots of men stop by her door for a chat, a smoke, and a cup of tea. She jokes with them and makes them laugh. She has the same kind of friendly relationships with policemen from the Tibbi station. They don't stop when they're on duty and in uniform, but they call in before or after going on a shift. A tall man with a large moustache has stopped and is leaning against the door. He asks me what I do, and I tell him that I'm a teacher in England. He questions Salma and leaves once he's satisfied that I'm not a new girl in the trade.

"He's a policeman," Salma states.

"Does he give you trouble?"

"No," she replies, rubbing her fingers together to indicate that she pays him cash. "All the police are good here if you give them money."

Victoria Unani Cream

Jamila looks terrible. Her business as a Tibbi Gali pimp is not going well, and she hasn't enough money to feed herself, never mind her cats. Her husband's leg is atrocious. Swollen and mottled purple with a great running sore, it looks as if it's in an advanced stage of decomposition. Mehmood has a plastic bag full of medicines hanging on a nail on the wall. It's a real potpourri—penicillin, serious-looking vials, and jumbo-sized tablets. He doesn't know what he should be taking, when he should be taking it, or how long he has been on this regimen. He's probably had such a mixed-up, irregular cocktail that nothing works on him anymore. He whips out some empty tubs of a miracle cream that he swears by.

"I put it on my leg to stop the itching. I put it everywhere," he says waving it at me. It comes in a pretty little pink, blue, and red box and its name is Victoria Unani Cream. The English words on the box are not enlightening or encouraging. They read: "For the protection of face frong small ball scabies and ring. Warm use Victoria Cream orfoot bum due to cold or after shaving and befor make-up use Victoria Cream every morning and evening. It clean soft and smooth your face." It says nothing about rotting legs.

Good doctors are expensive, so most of the poor of the *mohalla* never visit them. Instead they go to *hakeems*, traditional healers, or to people who set themselves up as medics without having any official training. The type of medic in most frequent demand is the *dai*—the traditional midwife. In Heera Mandi these women deliver children, and almost all have a quiet but busy sideline as abortionists. Many people, like Mehmood, don't even bother visiting the professionals. They simply buy likely sounding drugs from the pharmacy and hope they work.

A Marriage Proposal

I haven't been to Tibbi Gali for a while. Salma isn't here; she's gone to the shop to buy a cigarette. She's left a friend minding her room just in case a *tamash been* should appear. The plump friend sits on the doorstep. She has a big round face that crumples in around her mouth and folds into a toothless smile. And now here comes Salma, whizzing down the street like a little wild ball. She's shrieking and sliding on the wet stones. She pulls me into the room and shouts at me for not visiting her for so long.

Her friend sits quietly looking out into the street. She doesn't talk to me but she's not unfriendly. She puffs away on her cigarette and occasionally turns her face toward the dark interior of the room and smiles. Salma wants to take me to see a friend, and she shouts *"Chale? Chale?"* — Let's go? Let's go? — very loudly and very often as if I can't understand. The friend with the collapsing face is left in charge of business again and we walk a short way up the *gali*, around a corner, into a narrow alley, and then up a dark, steep flight of steps into a labyrinthine house.

An affluent man meets us on the second floor. He has the gloss and the ballast of the well-fed: he wears a thick gold chain and he's dressed in a crisp, beautifully ironed *shalwaar kameez*. I'm invited into a windowless room that has big locks on both the inside and outside of the door. The floor is covered in carpets and a mattress, and there are cushions against the wall. The roof seems a safer place to be, so we climb higher, the fat man breathless with the effort.

We are five storeys high and immediately below us to one side there's an enormous hole in the ground. The large interconnected building that used to stand there fell in on itself a few years ago. To-day only a rubbish-strewn hole remains where the basement had been. It's been turned into a giant open-air toilet and landfill site.

The fat man is clearly of some importance. He says he owns the entire building we've just climbed through and he waves his hand over toward the adjoining building. "That's my brother's house," he says. A young man scrambles over the roofs, jumping from one to the other and climbing over small walls. "My nephew," the man ex-plains and orders food and drink to be brought.

Salma sits on a *charpoy* drinking Coke and telling the man about me. He asks some questions about what I am doing in Heera Mandi and then his wife arrives. She's a small, young, stocky woman dressed in gold embroidered evening clothes even though it's four in the afternoon. She's wearing heavy makeup and has a confident, managerial manner. She and her husband complement each other extremely well. She's not his only wife — he has three — and I wonder how he has the energy, considering the stairs were such an effort.

The group discusses my appearance and the wife asks if I know about the business. "Of course," her husband replies. She weighs me up. She's canny, a businesswoman whose trade is women. They must run a protection racket or own some of the buildings in Tibbi Gali. I think Salma rents her room from them and they take a cut of her earnings. I get the sensation I'm being presented for approval.

Back in Salma's room it's going dark and she hasn't lit the candle. She lies on the blanket as the *tamaśh been* saunter by looking in. Two men stop and peer at us. They begin a conversation with Salma and come into the room. "Hello," one says to me in English. "I love you." I make to leave but they are closing the door and I'm instantly terrified in the dark. I can barely see and, as one of the men takes my arms, I stumble over Salma who is preparing for business. The man insists that he loves me and pulls at my arm while I edge toward the door, laughing nervously and wriggling to loosen his grip on me. Salma shouts at him to let me go and I push the customer barring the door. Outside his friend is still holding my hands, declaring his love and promising that we'll be married immediately. I'm running away, my heart racing. Everyone in the *gali* is looking and I feel such a fool. This encounter with the *tamaśh been* was too close, too dangerous, and I'm annoyed with myself. What else can I expect if I sit without my *dupatta* as night falls in a Tibbi Gali brothel?

Child Bride of a Monsoon Wedding

(Monsoon: July–August 2001)

The intensity of the June heat has gone
but it's been replaced by an oppressive
humidity that's inescapable outside air-
conditioned offices, luxury hotels, and the
homes of the rich. It's not all bad: Lahori
summers also offer compensations. It's a
sleepy city during the day and those who
are not working are languorous. Lahoris remain awake long into the relative cool of
summer nights: the restaurants are open,
and at two in the morning there are still
crowds around Data Dabar.

Heat and humidity keep the courtyard
empty on July days. The bamboo blinds are
lowered and the shutters of the houses are
half-closed to keep out the sun. Inside people sleep or doze, lolling on *charpoys* or on
the floor, their clothes crumpled and a sheen
of perspiration on their skin. The edge of
the road from Tarranum Chowk to Heera

Mandi Chowk has turned to mud that coats my sandals and leaves a gray-brown crust on the bottom of my *shalwaar*. Up in Tibbi Gali business is bad. The policeman hanging around Salma's doorway agrees. So do the rest of the women in the lane. They blame the heat. The men are sleepy—too tired for sex. Nazia has gone and nobody knows where. Perhaps business was bad for her too.

Many of Heera Mandi's residents find relief from the oppressive heat by sleeping on their roof terraces. As dawn breaks and light creeps over the rooftops, the village family dozes on *charpoys* in the open air. The family enjoys its comforts. The mother has a large fan wafting a breeze over her and the father has gone one better: he's brought their new air-conditioning unit up from the living room and has positioned it beside his own *charpoy* so that the breeze makes the edges of the sheet flap around his face.

Down in the courtyard a couple of rickshaws stop close to doorways. Their passengers are quiet, veiled, fast-moving women returning from all-night functions outside the *mohalla*. In the half-light of dawn dozens of rickshaws pass through Bhati and Taxali gates on their way to Heera Mandi. They see the *mohalla* at its very best: the early sunlight is soft and kind, and the sweepers have cleaned the courtyard and the *galis* and *chowks*. The Badshahi Masjid is surreally beautiful: its red sandstone walls and white marble domes glow in the morning sun. For a brief moment each day Heera Mandi shimmers.

Virginity

Ariba flings open the door and greets me jiggling with excitement. She looks older. Her hair has been cut and washed. She has a fringe and she's grown an inch or so. She's put on weight: I could've been away for a year rather than three months. Maha isn't at home and neither is Nena or Mutazar. They've gone to see a promoter. "Nena is going to London to dance," Nisha says in a great rush, so excited she arches her curved spine to stand straight for a moment.

Maha is away a long time so we wait and sleep on the mattress: Sofiya, Nisha, and I. Ariba wouldn't stay. She had something to do and vanished into the wet heat of the afternoon.

Maha and Nena are bubbling over with enthusiasm when they return. It's true: Nena is going to London in September to dance in a group. She'll dance in shows and Maha confides that she will have "relations" with men too. Payment for this means even more money— the promoter will take half the fee and Nena will have the rest.

Nena has had her official glossy photos taken. She looks sophisticated in a black dress and too much jewelry. I glance at her sitting beside me: she's prettier in real life without all the paint, when you can see the girl she really is. She's thrilled by the talk of London and glamorous dancing shows. Maha says that Nena will have to start her career very soon.

"She's going to Dubai to get married—as soon as we get the papers."

Nena nods, unfazed by the prospect. She knows what this involves: all Heera Mandi girls do, it's what they've been trained for since birth. There are few formal instructions; they absorb lessons from everyday life.

The papers Maha is talking about are a false passport and visa, ones that will make Nena appear to be 18 and the wife of a respectable man, not a child from Heera Mandi. When she arrives in Dubai the agent who is making the arrangements will marry Nena to a wealthy client. The marriage won't be legal and will probably only last a night or two, after which Nena will return to Heera Mandi with half the fee and without her virginity. It could be lucrative. Far better to do this, Maha explains, than to send her to England where her virginity won't command a price nearly as high as in the rich Gulf states.

In societies in which female sexuality is closely controlled, virginity is an important marker of ownership and a valuable commodity. In taking a girl's virginity outside marriage, a man deprives a girl of a husband and a family of its honor. Deflowering girls is connected with power and status because only rich men can afford to buy virgins—like Nena—girls who have no choice but to spend the rest of their lives as prostitutes.

Maha began the negotiations a few weeks ago and now the process seems unstoppable. The *dalals* agree. They say that Maha has committed herself to the deal: if she backs out, she'll have to refund the money the organizers have already paid. That's ten thousand rupees ($169) for providing Nena's false passport; forty-five thousand ($759) for arranging her visa, and another twenty-five thousand ($421) for the airfare: a total of eighty thousand rupees—more than $1,300.

"Can't she marry? I mean, can't she make a good marriage?" I ask, already knowing the answer.

Maha shifts uncomfortably among the cushions. She knows we're speaking from the same old script we have rehearsed so many times before.

"The daughter of a *tawaif* is always a *tawaif*," she sighs. "Marriage costs lots of money and we don't have any. And even if we did, who would want her? She's a *kanjri*."

Ariba has returned from her walk in the murky afternoon and is curled up on the back-crippling sofa. Maha hisses about a rape.

I think I've heard incorrectly and ask her to repeat it.

"Ariba was raped—just after you left last time," Maha says more loudly. Ariba stiffens and becomes motionless. "It was some *badmash*"—a gangster—"He took her into a house near the bazaar and raped her."

Ariba springs up from the sofa and dashes out of the room.

"We took her to the hospital and they said her virginity was gone. She didn't get pregnant though because her periods hadn't started."

Ariba is standing in the shadows of the other room. She thinks I can't see her, but I can make out the whites of her eyes as she stares at me trying to gauge my reaction. She's scared. Everyone is blaming Ariba for the rape. It's true she has been courting danger, and to prove the point, two of her uncles arrive at that moment to berate Maha for Ariba's loose conduct in public. There's a loud argument in Punjabi. They say that the family will be shamed if Ariba doesn't wear a *dupatta* in the street and spends so much time away from home.

The uncles don't know about the rape. Hardly anyone does. Maha wants it hushed up because a raped girl is bad for the family: it shows that they can't protect their women; that they have little social standing; and that they're not respectable. It's worse for the victim because once a woman, or a girl—or a boy—is known as the target of a rape she becomes so despised, so shamed, so worthless that she turns into public property. No one is raped only once.

Eating Their Sisters

Five well-fed men are sitting on a *charpoy* in the courtyard. They're young—in their twenties and thirties—and tending to fat, with fine, thick moustaches and glossy black beards. Mushtaq is with them, eating and smoking. He's put on weight and doesn't have quite the same physical charm he used to, but he still looks gorgeous as he wipes the mango juice from his beard. He's holding forth and the other men are laughing and attentive.

Maha and I watch them from the balcony surrounded by her plants. She has a rose bush with two brown-tinged yellow flowers. The third—and the best—was picked and given to me as a present. I have a seat on a blue plastic water tub, and I shuffle uncomfortably on the lid while trying to avoid sitting on the handle. In the midst of a profusion of dusty leaves Maha sits and fumes. "See the sister-fuckers. They keep looking at us." Looking is a method of control, and we'll have to go inside; otherwise we will be jeopardizing our honor.

"Look," she says again. "Look at them sitting there and eating their sisters." Eating food, she means, bought by the sale of their sisters' bodies.

Inside Ariba is skipping around and moves close to her mother when we sit on the mattress, smiling up at her. Maha laughs and pushes her away. Now Ariba wants to go to Dubai too. She says she'll go and earn lots of money and then they can buy a nice house. After all her beatings, after the constant, informal exclusion and all the years without affection, Ariba still wants to please her mother.

The Riches of the Gulf

The Best Musical Group, the VIP Musical Group, and a dozen other promoters have offices on Chaitram Road, a minute's walk from Tarranum Chowk. Young men sit at a table or at a desk inside a tiny room, its walls covered in posters advertising dancing shows in which brightly dressed women perform wearing a lot of lipstick and eyeliner.

These promoters recruit women from Heera Mandi, place them on their books, and call them when there's work. The promoters organize shows and make the women available for sexual services. The women are employed on short-term contracts, sometimes for an evening, sometimes for a week, and, at other times, for the luckiest, for a three-month tour. Some of the bigger promoters take women to the Gulf, and the most prestigious take their girls to England. It can be the chance of a lifetime. If they're beautiful and can please the customers, they can make a lot of money. When they return they can buy a degree of social mobility: VCRs, large fridges, and the jealousy of their neighbors.

Some bad promoters operate in the *mohalla:* they lie and cheat, taking women to the Gulf, then confiscating their passports and presenting them with a long line of customers. It happens often but no one talks about it: few want to risk their reputations by describing the ordeal they endured in a foreign country or the fact that they only received a fraction of the money that the many *tamash been* paid for them.

Two girls from the village family are still working in the Gulf. The girls are not especially pretty, and they're certainly not sophisticated, but they must be extremely skilled at the kind of services the clients demand. The eldest is on her fourth trip. She goes for three months, returns for a few weeks, and then departs again. All this coming and going is bringing in good money and there have been big changes in the interior of the house. A whole range of new furniture and electrical goods is stacked into the apartment; an enormous bed and dressing table fill the main room. A carpet has replaced the old rugs. It can't have been cleaned since it was laid, and mashed food is working its way into the weave. In a few months it'll build to a waxy, glossy shine.

The family is very attentive. They ask me to sit down and ply me with drinks. They're always friendly and the sons keep out of trouble and chat politely in the courtyard. During Muharram the family paid for a nice feast: they had cooks prepare four *deg* full of rice and chicken. Despite this, they still have not been integrated into the courtyard community. The others laugh about them, and I understand why they're called the village people. Two large hens are pecking and scratching around in the room, leaving droppings on the carpet and jumping on the new furniture. It's as if the family has never left its farm.

The hens stop clucking and look as if they are about to roost on top of the big new television. The family wants to show me something special, so they shoo the birds out of the room and bring in a DVD player still in its box. "It's from Dubai," the mother explains. "My daughter bought it." If there'd been enough space she would have swaggered.

The *gola walla* is offering a special today, and Sofiya and Nisha are pestering their mother for money. Mutazar whines too: he wants money for a giant balloon. Ariba lowers a few notes into the courtyard in a red plastic basket and the peddler ties a balloon like a six-foot condom onto the ropes. The *gola walla* balances five plastic bowls precariously in the basket. Although Ariba hauls it back up with slow care, the balloon comes loose and floats off up the *gali* and the basket swings violently so that the contents of the bowls begin to slop over the sides and the whole basket looks as if it's about to flip over. The special is a sludgy mixture of bright red syrup, ice shavings, and what looks like a tangle of pasta worms. The children say it's delicious. I'm not prepared to verify this.

The Bangladeshis who have a reputation for currying pigeons have moved. I spot the pretty daughter while Ariba is grappling with the basket. She's made up like a doll. The family's new house is just around the corner, its entrance on a side street only ten meters from the main road where there are a few shops, a gas station, and a tea shop. It's always busy: lots of men mill around and the family must be able to pick up a sizable amount of passing trade. I watch the father standing at the entrance to the *gali*. He's wearing a vest and a *dhoti* and is chain-smoking. He chats with the men, drinks cups of tea, and frequently meanders into the house with some of his newfound friends. The *tamash been* emerge after five minutes, by which time Dad is already back at the end of the *gali* scouting for yet more clients.

T he next day the Bangladeshi family is sitting outside the shops opposite Roshnai Gate. The father is annoyed that I haven't been to his new house, and I promise that I'll come to visit soon. His pretty daughter sits next to me and sips Pepsi loudly through a gnawed straw. She tells me her marriage is over. The man who made bedsheets was no good. She's adamant: "I'll never get married again." He was mean, she explains: he didn't give her money but still expected her to be faithful and not take any clients.

She huffs and adds, "I'm going to Dubai soon."

Her father remarks on the expense of the visa as his daughter looks directly at me. Her expression is chilling. Her eyes are not those of a child: they're flat and dead.

A Baby Sweeper

There's a new, much-loved baby named Hanouk at the sweeper's house. He's tiny, just 19 days old, covered in fine black hair and with thin, crinkly-skinned legs. He weighs nothing and lies in my arms with his big, *surma*-rimmed eyes staring intently at my face. He spends most of his time being handed from one member of the family to another. The trouble is, the children caring for him are often toddlers themselves. The new baby dangles upside down, his head skimming the floor until the young caretaker realizes this is wrong and jerks Hanouk upward, his tiny neck stretched by the sudden movement and his head and body at right angles. They take him into the other room, narrowly missing smashing his head into the wooden door. I wonder if he'll survive until my next visit. Tariq and his wife have three more children—all under five and all without obvious disabilities—so I expect that he will.

Auditions

A prestigious promoter is visiting Maha's house tonight. Her origins are in Heera Mandi, but Maha says that she lives now in a gorgeous house in Garden Town. The woman's name is Laila and she wears stunningly soft, embroidered silk. Her nails are an inch and a half long, square at the ends, and painted white. She's in her midthirties, and although she's not beautiful, she's so well-groomed, so expensively dressed, and has such an air of confidence that her presence in the midst of Maha's messy shambles of a home momentarily takes my breath away. She's accompanied by a middle-aged man—for protection and respectability—but Laila is clearly the one in charge.

They've come to see Nena dance. Laila is the promoter who is taking Nena to London, and she wants to take a second look at the goods before she pays any advance. Laila waits for the audition to start. She sits on the edge of the mattress and makes a studied effort to ignore a commotion of flying clothes in the other room. Nena and Nisha are fighting over ownership of a shiny yellow *kameez* that's topped with a black net blouse. The highly prized blouse is made of thin, brittle material. It's more like a plastic fishing net than a piece of fabric, with tears and holes and ragged edges, but it's the focus of a vicious power struggle between the sisters. Nisha claims it as her own. Nena, pumped up with confidence about her new career, demands the right to wear it. Nena insists that she really needs the black net blouse, unlike Nisha who has no chance of a dancing tour in London.

Nena wins the fight, even though she doesn't need special clothes; she'd look wonderful if she danced in a sack. She has a natural grace and Laila is impressed. When Nena has finished her impeccable routine, Maha suggests that she should dance too. Only Maha can't dance any longer—not the way she danced last year or even three months ago. She must have put on another thirty pounds in the past year. She can barely execute half the moves. Instead she relies on her facial expressions and the delicate and suggestive use of her hands. When she falls to her knees she struggles to stand up again; when she twirls around it is with a heavy step. It's embarrassing to watch, and I want to ask her to stop—but that would be an even greater humiliation for her. Dancing and being beautiful define Maha: they're what she has done since she was 12 years old. So I keep on smiling and when she's finished, drenched in sweat, we all clap and Laila says that she'll arrange work for Maha too. She's lying. She wants to flatter Maha. She's shrewd, and she sees Maha's desperation and her weak spots. Laila wants to get a good deal on Maha's daughter.

The Pimps' Turf War

I'm walking home through the dark, busy *galis* at ten at night. I've become accustomed to finding my way around in the gloom and gain my bearings by using special landmarks. There's a familiar, potholed stretch of road; a particularly offensive drain; a little kiosk selling corn fries in murky plastic bags; two old ladies spitting *paan* from the same worn doorstep; a sadistically unfriendly dog; and a bony, smiling youth always sitting nervously in an agent's office under a harsh battery of fluorescent lights. I know I'm nearing home when I see the barbeque stall run by a jolly man who splatters more oil over his belly than his chicken pieces.

I stop at an open door. It's an embroidery workshop specializing in heavily worked bridal *shalwaar kameez*. Boys of 9 or 10 sit on the floor threading beads and golden sequins onto rich red fabric stretched tightly over wooden frames. It's meticulous work and their little fingers work slowly and carefully. They knit their eyebrows together with deep concentration and the fluorescent light blanches their skins. One looks up and gives me an impish smile. "Hello. You America?" he shouts.

I never like the last part of the walk where I have to pass the men sitting on the *charpoys* in the courtyard. I know some of them and often they call out to me. I recognize some despite the darkness, pimps or drug pushers like Mushtaq. Others are unfamiliar faces: friends or clients of the pimps. Tonight the courtyard is unusually empty, and as I enter the *gali*, a volley of bullets is fired from Mushtaq's door, peppering the walls of the houses. The hierarchy of the pimps and the pushers is being challenged. The music in the courtyard has stopped. The popcorn *walla* has switched off his chimes and abandoned his cart. For once there's nothing but silence from Maha's home.

Thirty-six hours later the odd spent bullet is still rolling around the courtyard, kicked by the boys and collected by the younger children. The rest have been brushed up by the sweepers on their morning duty. Mushtaq stands, tall and shirtless, in the midst of a group of men. There's no question who won the battle.

Dog Woman

Maha is incensed. Laila, the silken-clad promoter, is, in Maha's opinion, a "fraud, a bitch, a *kusi*, and a *gandi* woman." Maha and Nena visited her in Garden Town yesterday to make arrangements for the advance on Nena's earnings. Nothing was arranged. Maha screws up her face and breathes out fast and heavily through bared teeth, "She steals girls and sells them in England." Maha thinks this is why she has a lovely home with air-conditioning in Garden Town. This is why she has a great big television, a DVD player, nice clean carpets, a car, and a husband who drives an even bigger car.

Laila has promised that Nena will be married in London. She knows a man ready to do the deed. He's young and handsome, and runs a hotel. Maha has seen a picture of him. He was wearing shorts so he must be a healthy, modern, and exciting sort of boy. But there will be no big one-off payment for the privilege of deflowering Nena. Laila makes it sound like she and the man in shorts will be doing Nena and her mother a favor. All the promoter is offering is a flat rate of one and a half *lakh* for three months' work plus half of the fee for any extra services Nena provides.

"Dog woman," Maha seethes. I'm tempted to agree. Laila is offering Maha so little because she knows she's desperate.

"My daughter is a *sharif* girl. She won't go to England for her marriage and get no money. She'll go to Dubai instead." Maha is adamant. And she adds, "She has an *ashik* here too now." A lover—Maha means a man who wants to have sex with Nena, a man they will later describe as her *shohar*—her husband. There are many words for men like this in Heera Mandi: *ashik, mehbub, sanam*—lover, beloved.

Maha shakes my arm to stress the point, and she has a very powerful grip. "We met him at Laila's house. He's a friend of Laila's husband." She's working herself into a fit of indignation. "Does she think my daughter has no honor? Does she think I have no honor?" She nods her head furiously, and a man who is sitting quietly in the corner smiles politely.

"Who is he?" I whisper.

"He's a good man. He's the *muezzim*. He calls the *azan* in the mosque."

The man adjusts his crocheted orange *topi* and smiles some more.

He has a big black beard and a humble manner. I wonder what he thinks of the conversation and if he's uncomfortable about being in the home of a *kanjri*. I expect he's too poor to be fussy: he's looking for work as a Quran teacher and needs to humor potential patrons. He is given a plate of rice and chicken and Maha asks him to run a few errands. He's sent to buy *sita*—roasted corn cobs, lots of them, that he brings back in two bulging shopping bags. There's one for each of the children, one for him, and two for Maha. She rummages through the bag, inspecting the corn and choosing three of the tenderest and juiciest for me. "Why three?" I ask and she replies that she knows how much I like them. And then she adds that I'm looking thin and that it's not a good state to be in because how will I ever find a decent husband?

An Impatient Landlord

I'm sitting on a *charpoy* near the pimp's den, drinking Coke with one of Maha's cousins. It's early evening and a drama unfolding in the corner of the courtyard is generating an enthralled audience. Maha's sister is being taken to hospital. Six people are squeezing into the taxi. Most of them are obese, and all I can see through the windows of the car are squashed bodies and contorted faces. Someone says Maha's sister is dying and a rumor of her imminent death begins to spread.

The *charpoy* is a good vantage point and the courtyard looks different from this angle. It's still hot and humid, but the sun has disappeared behind the houses and we're in shade. Only a few of the pimps are around. Maha is watching me from her balcony, gesturing angrily and questioning why I'm sitting here in the open. Five minutes later she's rushing toward me. She's changed into one of her best outfits. It's a cream chiffon creation with a gold-sequined neckline. It's entirely inappropriate for sitting on a dirty *charpoy*, but Maha likes to make an impressive public appearance.

Maha's cousin is a pimp. He works as a minder for groups going to the Gulf and he has dozens of photographs to show us. Some are of groups from Mumbai and Heera Mandi. A few of the Indian girls look emaciated, and a woman peering over my shoulder says that one must be an AIDS case. Others are very overweight. One of Maha's fat cousins is a singer in a group and she features prominently in most of the shots.

A girl from Heera Mandi is gorgeous. She has a delicate face and I point her out several times and remark on her beauty. Maha laughs and says I'm a mental case. The girl isn't nice. There's nothing nice about her. She looks like a dog.

The clubs in which the women are working are poor and low-class, furnished with white Formica tables and metal-framed chairs with plastic, leather-look seats. The curtains are cheap and badly hung and the carpets are fraying around the edges and creeping up the walls. These women are not entertaining rich Arab sheikhs when they go to Dubai. They're servicing the Gulf's migrant workers: men from their own country, many of whom will be lonely and frustrated and who will have saved to enjoy the pleasure these women offer.

The cousin had lots of family photos taken in Heera Mandi around twenty years ago. Maha is photographed as a teenager, looking very pretty alongside Fouzia, her equally pretty sister. And then there are her cousins, who were blessed with none of the pretty genes. A black-and-white snapshot was taken of Maha when she was about 8, and I'm struck by her similarity to Ariba.

"Sofiya looks like I did, doesn't she?" Maha comments. She does in a vague kind of way, but Maha cannot—or will not—see that her least favorite child is the one who looks most like her. And then, finally, there are photos of me. Some were taken at a party and others at a wedding. I'm shocked and have to go through them again. I look so out of place—I hadn't realized how badly until I saw these pictures. I always thought I fitted in and merged with the surroundings—as if a *dupatta* made me look like all the people I see around me. But I don't. I look so weirdly, disturbingly white.

Maha stirs nervously on the *charpoy*. The *mallik* has arrived for the rent again and she hasn't any money. He rides up on his motorbike to humiliate Maha publicly for nonpayment. He must treasure his bike because he's taped layers of bubble wrap around the suspension and gas tank. Brown synthetic fur is tied along the long arms of the handlebar mirrors. They look like antlers. I can't decide whether this is for protection or ornament.

We dance in the evening to forget the *mallik* and his threats to turf the family out. Even Ariba is dancing. She's never done it before, even in play. She's a bit wild and full of energy, but Maha promises that she will pay Master Jee to tame her and channel her exuberance. When she stamps her feet she will learn to do so with grace. We practice with the music so loud that they will be able to hear it in the bazaar, and when Maha throws open her shutters, the restaurant customers on Iqbal's roof terrace gather to watch the dancing girls of Heera Mandi perform for free.

Drugs

In the winter it takes fifteen minutes to walk to Bhati Gate to buy my newspaper; in the summer heat it takes a slow twenty minutes. I like getting out of Heera Mandi. I like the bustle of the roads in the old city and the tobacco shops near the gate. I walk slowly here, savoring the smell of the ropes of tobacco stacked in tall piles and the smoke from the *hookahs* of the men who sit in their vests in the shade of the shops.

By contrast, the stench is so vile in the butchers' section of the road that it's enough to make me retch. A chicken is being killed in the correct Muslim way as I hurry by: its throat has been cut, and it will die in the butcher's blood-encrusted, blue plastic shaking barrel as its heart pumps the blood and the life out of its body. Another butcher is slicing lungs. He has an unusual technique: he grips a knife between his toes so that his hands are free to maneuver the offal and to place it into neat mounds. In Tarranum Chowk the toes of today's fresh corpse have also found a new use. Passers-by have been sticking ten-rupee (seventeen-cent) notes between them so that someone can organize a funeral. The notes flutter in the breeze like two tattered fans.

T wo stigmatized communities live side by side in Heera Mandi: the prostitutes and the drug users. The addicts gravitate toward Heera Mandi because no one is going to get annoyed with them for lowering the tone of the area. Most of the addicts are from out of town. They began as unskilled laborers living in hostels or sharing small rooms, earning a pitiful living. Many must have been sad, isolated boys and men a long way from their homes. They still had enough money, though, to pay for some good times, and they bought drugs and enjoyed themselves with local women. They took hashish, and then heroin, until the addictions took over and they ended up on the streets, sleeping by the *masjid* or in the gardens or wherever they collapsed.

A few years ago, when I first came to Lahore, the drug of choice was heroin. The addicts smoked it on street corners, huddling together in little groups. Now they inject a cocktail of pharmaceutical drugs—tranquilizers, painkillers, antihistamines, and anything else they can lay their hands on—and grow festering abscesses in their arms, legs, and groins. Injecting is cheaper and the supply is more regular because they can buy the drugs from pharmacies. No prescription is ever necessary. The addicts share needles and syringes, and it's only a matter of time before HIV tears into this bedraggled group of men and the prostitutes who service the ones who have yet to topple into stupefaction.

It's difficult for me to engage with the area's visible drug culture, even if I wanted to. The women, by and large, don't shoot up or smoke heroin. They're like Maha: they take sleeping pills and drink cough medicine. Most women will also take hashish and *booti* when they can find it. But an entirely male-dominated supply network controls the harder drugs, like heroin and the injectable hospital-grade medicines. It's nearly impossible for the women, confined to their houses, to access these networks without the help of a man. For a woman to be seen taking drugs—even a cigarette—proves that they are *kharab* and *beysharam*—spoiled and shameless. Sending someone to buy pills or cough syrup from the drugstore is a lot more respectable.

A supply depot operates from a shop at the corner of the road, only a few meters from Iqbal's house. A couple of dealers hang around the pool table outside. I wouldn't dare to investigate the underground room into which men disappear, or even to look in for too long as I walk by. It's part of the men's world from which I'm excluded. The person in charge of this drug dealership is the patron of the splendid *sabil* that's erected in the street on every important religious holiday, but he's only a small player in the drug business, a client of others who are much wealthier and more powerful. Like Maha's madam, the lovely wife of the army major, they live in a far superior part of town.

Rats

More rats than people live in the old city. They are everywhere and I've grown accustomed to gray shapes scurrying along walls, slipping in and out of drains, and disappearing into cracks in the brickwork. As we eat lunch on Maha's mattress, three small rats are watching us from on top of the curtains covering a couple of nailed-up doors. Maha laughs at them and treats them like pets, but she isn't always so amused. She pulls down her *kameez* to show me a chest crisscrossed with encrusted scratches. A rat the size of a small cat ran over her a few nights ago and they got into a fight. When Adnan last visited, he woke up screaming, a giant rat gnawing at his toe. He'd drugged himself into unconsciousness and the rat must have thought that he'd died.

I take the dishes into the kitchen after the meal and my stomach turns: two rats are eating out of the pan from which Nisha has just served the vegetables. The leftovers will be reheated for this evening's meal, and I make a mental note to refuse supper. They're so bold that they return to sniff around my feet and snatch bits of food from the floor while I'm doing the washing up.

If I were a rat I'd move into Maha's house. It's ideal, with lots of lovely secret places to make a cosy nest—and friendly neighbors and a constant supply of food lying around. This is going to change. I've been investigating the housekeeping in detail, and Maha is persuaded: we are going to launch a cleanup operation.

The main stairwell of the house runs right through the center of the building. It's been blocked off for years so that the house could be divided and sublet more easily. Only the rear, narrow, spiraling steps are in use. Old furniture, clothes, shoes, and bedding have accumulated on the main stairs and rubbish has been chucked on top and left to fester. I think there are several families of rats living among the debris because it stirs every so often; there's a swiftly moving wrinkle in a piece of crumpled fabric and a sheet of paper rustles and quivers. A little room at the front of the house is off-limits to Maha. The *mallik* keeps it for furniture storage. It's full of old mattresses and chairs. I prise open the door far enough to see inside, and two enormous rats dash into a hole in a large mildew-speckled sofa.

We are going to sort out the stench as well as the rats. One of Maha's rooms is barely used. It's stacked with useless bits of furniture, old burnt pans, and lots of plastic tubs containing water to see us through the periodic interruptions to the supply. A toilet was added to the room long after the house was originally built. It's a bad design and gets blocked easily. Today it's chock full of shit and hasn't been cleaned for weeks.

"That's Ariba's toilet," Maha explains. "She's such a *gandi* girl."

The Cleanup

Tariq, the sweeper, has agreed to get the cleaning off to a speedy start. This morning he and a gangly youth came over to Maha's after they'd finished the street sweeping. They've left their cart in the *gali* and are standing in Maha's best room assessing the task ahead. They've brought brushes, buckets, and a bottle of acid that Tariq promises will shift anything.

Ariba's toilet needs something awesomely powerful. The shit has been fermenting in the heat and threatens to bubble over on to the floor. Tariq is hardened to vile odors and other people's filth, but even he looks alarmed. I can hear Maha explaining that she never uses that toilet, she uses the one in the best room. The mess isn't hers, it's Ariba's. She's the dirty one of the family.

Tariq pours the acid into the brown froth and sprinkles more over the floor. A lot of loud fizzing and a steamy cloud of sulfurous gas fills the bathroom. The toilet's contents are boiling and then gradually draining away. In the other room we are all choked with the gas, and Tariq is coughing loudly and fighting for breath as he wields his brush and whips a volatile mix of acid, urine, and feces around the toilet floor.

Tariq's acid is good stuff. It strips the toilet bowl and the floor of its topmost layers and the place looks passably clean. It has probably scorched his lungs too, but he doesn't complain. You don't if you're a sweeper.

The gangly youth is in charge of moving rubbish down the stairs and into the cart. He's already been on two runs to the dump at the end of Fort Road, and he'll be making several more journeys. Debris is everywhere: bits of *roti* and *naan* bread, mango stones so old they are black, a pair of Sofiya's underpants still containing a smelly accident, wet shoes left to rot. Tariq is cleaning the floors, tipping buckets of water onto the concrete and brushing with vigor. Maha and children are crowding around to watch the performance. They have never seen anything quite as efficient.

T he worst of the stench has gone by evening. Tariq and the youth have hauled five cartloads of rubbish to the dump. They've also wrapped two giant rats in *shoppers* and taken them outside, wriggling and bucking, so that they could crush their heads with a shovel. The other rats have abandoned the kitchen and Maha's best room for a while. They're retrenching and have taken refuge in the *mallik*'s furniture store: I peeped through the door and saw them carrying on the party in there.

Ariba is cooking something special. She's been sitting by the pot for hours, stirring and watching, adding bits of ingredients or a cup of water. She's intent on what she's doing and has been into the family room only two or three times. I know when she does without looking up because I catch a whiff of stale urine and dirty clothes as she passes the air cooler. Her yellow satin *shalwaar kameez* is filthy and she's worn it for days.

Two days later Maha's rats have seized the initiative. They have regained all their old confidence and are rushing in and out of the *mallik*'s furniture store. But we are not beaten yet. We are launching a counteroffensive and the cleaning operation is entering phase two. I've been to the up-market Al Fateh supermarket in Gulberg and have bought imported cleaning solutions. The Pakistani variety are a lot cheaper, but I want to stick with the products I know. I don't wish to encounter anything as potent as Tariq's acid.

I've put on a pair of rubber gloves and am sorting out the brushes. I'm quite excited. I've started on the finer points of Ariba's toilet and Maha is reeling from shock.

"My Louise, my Louise," she gasps. "My sister is doing sweepers' work." She talks about shame, but we have no meeting of minds on this topic. There is an irreconcilable culture clash. I consider her shamed for not cleaning her home. Maha thinks I'm shaming myself by cleaning toilets, and in a bizarre and inexplicable twist, I'm happy to be doing so.

I have an audience. Some of the local women are hanging around the doorway trying to catch a glimpse of me in action. I can hear them discussing the spectacle and calling to others. "The woman from London is cleaning the toilet."

A water shortage halts the washing of the clothes. Two giant buckets have been filled with soapy water and all the dirty clothes in the house are soaking. The *niche walla*—the man from downstairs—has turned off the supply. The landlord has given instructions to limit the supply to Maha and the Pathani woman upstairs. Water is scarce and it costs money.

Maha screeches from the balcony that we need water to rinse the clothes. The *niche walla* is grumpy; he saunters from his *charpoy* and switches the supply back on for a few minutes. Frantically we try to fill buckets and bowls, but it still isn't enough and we have to call the *mishar*—the water carrier—as a last resort. He has a skin full of water slung across his shoulders, so heavy he walks bent into a right angle. Each skin fills four or five buckets and the charge for his service is ten rupees (seventeen cents). I've seen him many times this summer trudging steadily back and forth across the courtyard selling his water to households with shortages. Today, his tread is heavy and his feet scuff the floor as he climbs the stairs. He stops halfway up to catch his breath. He's perspiring, his face creased with effort, and then, as he opens the skin and the water gushes into the buckets, he begins to uncurl. He will make three journeys to Maha's tonight.

Mota: The Fat Lover

There's only one subject to talk about this evening: Nena's *mehbub*—her lover—the man they met at Laila's. He's rung Maha twice and wants to see Nena. They say he's old, black, and ugly with a scrunched-up face that Maha mimics with a grotesque leer. They call him Mota—Fat—because he wobbles when he walks, and because he's a great big fat catch.

The monetary compensations are attractive even if his body is not. "He owns factories and he's rich," says Maha, shaking my arm. "If Nena marries him she won't have to go to Dubai. We can pay the *dalals* back for the passport and the visa. He's so rich he carries a *lakh* of rupees in his pockets. We've seen it," she explains. "Ask Nena."

Nena confirms it. "It's true. I saw a *lakh*." She almost swoons.

"He's got a big white car with air-conditioning. He wants to marry Nena and he'll give us one and a half *lakh*," Maha says and then pauses. "But I don't know."

"I want to get married to him," Nena states, trying to encourage her mother to agree. She smiles bashfully, casts her eyes down, and covers her mouth. "I want to marry him. I want him to take us shopping." She flops onto the mattress, buries her face in a cushion, and laughs. Her mother joins in. At last, they think, their luck is changing.

ota has telephoned Maha. They chat while a radiant and smiling Nena sits on the mattress. The *ashik* wants to speak to Nena, and she skips over to take the phone, beaming at it and self-consciously smoothing her hair. Maha keeps whispering very loudly and giving her instructions on suitable things to say. "Tell him you like him. Tell him you want to see him."

Nena does not need to be told. She laughs, she giggles, she tells Mota that she thinks he's nice. As she does so, Maha keeps reminding me what a rich man he is. "He's a good man. He's ugly and old, but he's good."

Nena grows more animated. She covers the phone and says something about presents. Maha does a little dance and, taking the phone, rounds off the conversation with the *ashik*. He's coming to the house in a couple of days to see Nena perform her classical pieces. A flushed Nena collapses on the bed.

"Louise Auntie, he's bringing me a present, a CD player like Laila's." As she speaks, she holds her hand over her mouth and her eyes sparkle. The other girls look on admiringly. Nena is thrilled — and I'm confused. I thought I was coming to Heera Mandi to document a terrible trade, and yet Nena is seemingly not being dragged into prostitution: at 14 she's embracing her family's business with enthusiasm. She's going to do what generations of girls in her family have done before her, and in twenty years' time she will be like her mother—abandoned and dependent upon the sale of her own teenage daughters to survive. But, for now, she's flattered that Mota wants to spend so much money on her. It's a reflection of her status as a beautiful, high-class dancing girl. Probably for the first time in her life, Nena is exercising a form of power, and she's enjoying it.

Monsoon Floods

I had planned to do some interviews in Tibbi Gali at lunchtime, but it was impossible to leave the house. Two hours of torrential rain kept me inside, watching the spectacle from my window. An addict lay on the wall running along one of the big old houses in the courtyard. Water from the roof coursed down on him in a powerful foaming chute, and he lay semiconscious in the torrent for half an hour before staggering off to find a drier spot to slumber. Young men in *dhotis* sat bare-chested on the *charpoys* and exhilarated boys fought and frolicked in the impromptu swimming pool, but after a while, even they grew tired of the rain and went inside.

After an hour the deluge began to drip through my ceiling. The courtyard was like a lake, and when the storm had passed, the roads of Heera Mandi were left in a stinking, muddy mess. Great puddles and ponds of rainwater, shit, rubbish, and sludge sent the pedestrians zigzagging up Fort Road. The storm had washed sewage up from the drains and into the streets, and a fearsome stench hung over the area as pedestrians hopped from one drier patch of earth to another. Near Roshnai Gate brown slurry spewed like a geyser from a manhole. I took a rickshaw to the post office on Mall Road in the center of town in order to send my children some letters, but I never managed to make it. The roads were several feet deep in water, and as the liquid sewage lapped around my sandals and the rickshaw engine died, I decided to leave the journey for another day.

Twenty-four hours later, Jamila's room is still three inches deep in runny mud. A dozen wet, stringy kittens have taken refuge on the *charpoy* and on the piles of rubbish at the back of the room. Jamila is more wizened than ever and Mehmood's festering leg is no better. It can't improve in this environment. The couple are covered in beads of sweat and steaming gently under the low-slung plastic roof.

Three big lumps of masonry block the road a little way down from Jamila's room, and a *tanga* is having trouble finding a way through. A couple of emaciated workmen are staring into a shop not knowing where to start. They face a considerable challenge: the interior is filled with rubble and the upper three storeys have collapsed, one on top of the other. "There was too much water," someone explains looking up through the chasm to the light above. Yesterday's rains were truly devastating throughout northern Pakistan. Rawalpindi had its heaviest rainfall for a century and hundreds are believed to have died. In Lahore we escaped lightly.

Preparing for Mota

Maha has slept off the worst of an overdose and is organizing the household. Mota is coming to visit tonight to present Nena with the much-trumpeted CD player. I arrive early to a flurry of kisses and am made to sit in the middle of the arrangements. The sheets are being changed and the mattress is now decked in red, black, and yellow geometric patterns. The kitchen looks equally frightening and a lot less clean.

Maha's house is an unlikely scene for expensive romance. Mota will have to drive the big, white, air-conditioned car up the filth-choked, open-drained, potholed roads because men as rich and as fat as Mota never walk anywhere. After he negotiates the litter-strewn steps, the great *ashik* will climb through two floors of narrow, spiraling darkness. If he doesn't turn back from exhaustion or fear, Mota will have to paddle through the corridor that serves as Maha's kitchen because the floor is inch-deep in water. He'll pick his way between the pots, pans, buckets, stoves, unwashed dishes, ladles, and the wire-mesh rack containing rotting onions, plates, and assorted dirty rags.

I can't imagine that Mota will be impressed by the decor, but I'm sure he will be enchanted with Nena. She's wearing a black silky *shalwaar* and a new red *kameez* embroidered around the neck and hem with gold thread. The outfit is too large for her and hangs loosely on her shoulders and gapes at the neck. It makes her look fragile and vulnerable and very, very young. He'll love it.

She's applied foundation, a lot of black eyeliner, and bright red lipstick that's the same shade as her dress. She rummages through her mother's jewelry bag and tries on everything twice, finally settling on some elaborate dangling earrings and a splendid artificial gold choker. Her hair is tied up and she spends twenty minutes curling the tendrils around her ears. She's looked in the mirror so many times that, if I didn't know better, I would believe she was preparing for a first, and longed-for, date with the man she loved.

Mota rings to say he will be delayed. He's at an important meeting at the Pearl Continental Hotel and is discussing the purchase of some vital new machinery for his factories with a group of engineers and salespeople. Nena and Nisha use the delay to practice poses culled from Indian films. An hour later Mota rings to say that the meeting is dragging on but that he will be with us soon. The girls have given up modeling themselves on film stars and are lying on the mattress getting their clothes badly crumpled. Adnan calls to tell Maha that he's dying and too ill to give her any money. She breathes deeply and controls the rising panic: she's still hoping that Mota is coming to rescue them. Another hour passes and the fat lover promises that he'll be here in fifteen minutes. We no longer believe him — and he doesn't come.

By half past ten the excitement has died down. We eat the food lovingly prepared for Mota, and Maha says that he's a sister-fucker like all the rest. Nena is feeling rejected and sits in a corner despondently hugging her knees. Picking up a little mirror, she pulls a face and sighs, "All my lovely makeup is spoiled."

My guess is that Mota is interested in Nena but doesn't want to come to Heera Mandi. It's not just because of the pimps and the toughs and the equally worrying police, but because of the stigma.

Rich men still buy girls from Heera Mandi, but they enjoy them in luxurious hotels, smart private apartments, and houses with gardens, indoor plumbing, and American-style kitchens. Like Laila's house. Laila has heard that Mota wants Nena, and she wants a cut of the fee charged for Nena's virginity. She's suggested that she can keep Nena safe in her nice house for three or four days for Mota's exclusive—and discreet—pleasure. For this service, Laila is demanding half the virginity fee. Maha refuses. Bringing Mota to Heera Mandi was Maha's doomed attempt to cut out the agents and the pimps. She failed miserably and now Nena is looking for another rich *ashik*.

Badmash

Someone runs behind me as I walk down Fort Road, draws level with me, and pulls at my sleeve. It's a man—and not anyone I know. I make to go into the bakery but he persists.

"Wait," he cries. "It's me. It's me."

I look, and then look again. It's Tasneem, the *khusra*. It's little wonder I didn't recognize her: Tasneem has become a boy. He's wearing men's clothes and his hair is ragged and short—far too short. It's been hacked off, and in places I can see his scalp and scabs from healing wounds.

"What happened?" I gasp.

He's embarrassed and keeps touching his head. "It was a *badmash*. He said he loved me, then he put a gun to my head." He reenacts the horror in the road, *tangas* and rickshaws veering around him, and then he grows calmer and we step into the shadow of a quiet *gali*.

"He said he would kill me and then he cut off my hair with a knife." He's crying and rubbing his eyes on his sleeve.

"Where are you living now?" I ask.

He shrugs. "Nowhere. Perhaps I'll go back to White Flower."

"Where do you sleep?"

"By the mosque."

"And what happens when it rains?" I ask, stupidly.

"I get wet."

Poor, poor Tasneem: his new love and his new life have come to nothing and now he's back here in the very place that he was so desperate to escape from.

"Come to see me, please," I ask. "You know where I live."

He promises he will come, but I know he won't. He kisses and hugs me, reminding me that I'm always his sister. Then he's gone, running and skipping between the rickshaws, waving, smiling, and crying at the same time.

The Fat Lover Comes Courting

Maha and Nena are not at home, and those who are left are very excited. Nena has been summoned to Mota's side: they are meeting now. A string of people turn up and sit with me while I wait for them to return. First there is Farrukh, the *dhobi walla* who also doubles as a *dabana walla*. Farrukh has been helping Maha around the house recently in return for meals. He gives me an exquisitely painful foot massage and Nisha says I shouldn't have let him: it's shameless. Then comes Master Jee, who smokes furiously and jumps from the chair and departs with a dramatic stage walk. The Pathani woman from upstairs calls in and strides about the room looking slightly aggressive while talking in a loud voice about her bad luck. Finally, I'm introduced to a middle-aged man. In a loud rasping whisper Nisha says he's the biggest *dalal* in Heera Mandi. Word must be getting round that Nena is about to make a good marriage and that the fortunes of the family are on the way up.

The bride-to-be and her mother return in an upbeat mood: Mota will definitely visit this evening. They have just spent two hours with him at Laila's house and then he dropped them at Babar Bakery near to Data Dabar, half a mile from Heera Mandi. Maha dumps a bag of shopping on the mattress.

"Idiot," she cries at Nena half-jokingly. "Look at what you got us." Maha tips out the contents of the bag. Mota had asked Nena what she would like from the bakery and food store. "You can have anything," he said, getting out his *lakh* of rupees. Maha stands with her hands on her hips and looks at the purchases.

"Guess what the idiot said: 'One bottle of Pepsi; two bottles of fresh water for Louise Auntie; and some jam.' Nena is going to marry a rich man and she asks him to buy her water and thirty-five-rupee [fifty-nine-cent] jam. I have a totally stupid daughter."

She's too happy to be cross for long, and she laughs and tells Nena to be sensible next time and ask for a lot of expensive things. The whole family is happy. Sofiya has kissed me at least a hundred times, and even Mutazar is being good.

Mota arrives after night has fallen. He pulls into the courtyard in his big white car, a brand-new Land Cruiser. No one has seen anything as grand and as expensive around here for a very long time.

His nickname does him justice. He's staggeringly unattractive and must be three times the size of Nena. The one thing they got wrong was his age: he's only about 45.

We are being introduced when Sofiya, who is supposed to be corralled in another room, dances in, twirling with excitement. She gasps and presses her hands to her cheeks and shouts, "Mota came. Mota is here. Mota. Mota. Mota." She's not being rude: she's simply saying what she understands to be his name. Ariba catches hold of her hair and drags her out. There's a look of absolute astonishment on the little girl's face. She wants to join in the party and drink Pepsi out of glasses that have their sticky labels left on to show how new and clean they are and how special the occasion really is. She doesn't know what she has done wrong. If Mota has noticed the insult, he doesn't show any sign. Perhaps he's too obtuse to see the link. Perhaps he doesn't care.

Nena charms Mota with an elaborate performance of hair-brushing. She has long slightly wavy black hair that reaches well below her waist and she swings it about, half-hiding her face. Maha has gone to get some snacks, and I'm left alone with Mota and Nena as a kind of chaperone. There's an uncomfortable silence, and Nena smiles continuously while fidgeting with her hair. Mota plays with his mobile phone, and I marvel at how such fat fingers can operate such a tiny device. It's a relief when Maha comes back in. She jokes, laughs, and flatters Mota who insists on calling her Mother while she coos and calls him Son.

Maha performs her social role perfectly. Like other successful prostitutes she's learned how to be an accomplished socializer who knows how to entertain men outside, as well as inside, the bedroom. Their livelihood—and sometimes their life—depends on the ability to read people and assess their mood. I've often wondered if this is why the prostitutes I have met have usually been friendlier and more approachable than most other women I know.

Mota doesn't stay long—which is good because there's always at least one power cut in the evening. It wouldn't have been wise to have Mota sitting in the dark, without a fan to cool him, while the rats run their familiar circuit around the room. Some of them are so big you can hear the thud of their feet.

Maha and Nena accompany him to the car. I watch from the balcony with the other children and the recently returned Master Jee as Mota maneuvers the Land Cruiser out of the courtyard, reversing and negotiating a path between the piles of rubbish. Everyone is doing wild dances, hugging each other and repeating over and over again that the car is so expensive and that Mota must be incredibly rich.

Nena is glowing and Maha is triumphant: Mota wants the marriage to go ahead soon, and it won't be just a three-day event. He wants Maha to keep Nena for his exclusive enjoyment whenever he visits Lahore. She's to entertain no other men—and she will be paid handsomely for her faithfulness. Maha sighs and looks at me and Master Jee. "My life is going to be good. We will have a new house, and we won't have to worry about the rent and the electricity. Maybe we will have a car. I am still alone: Adnan has gone, but, so what? My daughter is going to have a rich husband. Nothing else matters."

"We will go shopping soon for my wedding dress," Nena says

proudly. "Everyone will have a new dress. Louise Auntie, you will have a new dress too." I say that I have many dresses and don't need any more. "No," Nena replies softly but firmly. "It will be my gift to you for my wedding. I will ask my husband to buy you a dress. Please, it will be my gift."

She is sitting right next to me. I expected her to be appalled or frightened or resentful at the prospect of a marriage to Mota, but she's not: she's thrilled. She's giggling and embarrassed and slightly shocked, and she has something important she wants to tell me. She whispers so that no one else can hear.

"Louise Auntie, he kissed me—on the stairs, in the dark part. He went like this." She bends close to me and kisses me on the cheek. "And he held my hand and put it like this," she takes my hand and places it on her chest. "And he said, 'Give me your heart.'" It sounds beautifully sweet. If Mota was 15 rather than 45, and if he wasn't paying to buy Nena, the scene might have been endearingly romantic. Nena is certainly impressed by it. She repeats the same little story and actions five times in half an hour. Perhaps it's to make sure I've understood or maybe she wants to relive the exciting experience over and over again. Even so, a few doubts remain to mar her pleasure, and when her mother is carrying on about how honorable and good and rich Mota is, Nena grows sad for a moment and then suddenly leans toward me and whispers, "He's got tiny eyes."

m ota's wife is happy and today, a day after her husband's visit to Heera Mandi, she telephones Nena. The two of them are having a good chat. The wife asks how Nena is feeling. Nena replies that she's fine but a little nervous. The wife says not to worry: Mota is a nice man and everything will work out well. Nena asks after the health of Mota's beautiful children and smiles when she hears that they too are fine. Maha is sitting close by, nodding in satisfaction. There's no spoken hatred or jealousy: nothing except a formalized friendship between the fat lover's wife and his teenage girlfriend.

"Why isn't Mota's wife angry?" I ask. "Why does she want to talk to Nena and be her friend?"

"Look, Louise, Mota is a rich man," Maha says in surprise. "He can do what he wants. His wife has to keep him happy; otherwise he'll divorce her. If he's happy, she'll have a good life. If Mota is happy, we'll all be happy."

Mota is pleased that his wife and new girlfriend have established a good rapport, and he's rung to say a few nice words to Nena. She tries to sound relaxed, but she can't help squeaking: she's lying in her mother's lap having her eyebrows plucked. Maha has stretched the skin so tightly over her brow that the girl can barely move. Even in this agonizing position Nena looks beautiful. Two tiny new faces are looking down at the beauty treatment from the top of the curtain rail. A large family must be breeding in the *mallik*'s old sofa.

Lessons in Seduction

A few nights after Mota's visit, Nena is having a lesson in seduction. She's practicing her routines and Maha is adding a few new moves or encouraging her to exaggerate existing ones. "That's it, bend over a bit further. Now, move your hands like this." Maha demonstrates. Then they practice exactly how Nena should look at a man from behind her fingers and how she should reveal her cleavage as if by accident. Maha is teaching Nena how to be provocative. In a Western context these little glances, the direct eye contact, the shake of the breasts, and the sway of the buttocks would hardly register as sexual displays. Here in Heera Mandi it's the Pakistani equivalent of lap dancing.

"Master Jee can't teach her this," Maha laughs. She watches her daughter fall to the floor as if exhausted from intense but thwarted desire and adds in satisfaction, "Mota will die of a heart attack."

Nisha wants to dance too. She puts on a dress that Nena says is her own and the girls begin screeching at each other. Nisha does a jerky dance and Nena sits looking petulant. "This is my song," she announces. "You can't dance to my song." It's not like her to be so petty and mean. Maha tells her not to be stupid and Nena turns on us.

"You are all jealous of me because I'm marrying a rich man," she declares and flounces out of the room. She curls up in the corner of the bedroom. Ariba says she's crying. I've tried to talk to her, but she remains wrapped in a tight little ball and refuses to speak.

Sofiya has lightened the atmosphere. She wants to dance like her big sister. Even at 3 years old she's remarkably good. She has picked up more than the basics. She twirls, she raises her hands, she peeps through her fingers, and she sings along to tunes about her *mehbub* and her broken heart.

Everyone is enchanted by the performance. They are laughing and cheering and shouting, *"Bund chalo, bund chalo,"*—"Move your bum"—over and over again. Sofiya complies with perfect, wonderfully timed gyrations of her hips. If she continues to make this kind of progress and can maintain her style, she's going to grow into the kind of dancing girl of whom legends are made.

Black Magic

Maha is at her mother's house. She's high on something, swaying to and fro, struggling to speak. Her sister has returned from hospital. High blood pressure caused her recent health scare, so she's taking it easy, lying on the floor in a heavy, rosy-cheeked heap next to the husband of Maha's dead sister. They've been having a relationship ever since Fouzia died in childbirth. Maha doesn't approve. She thinks it's an insult to the cherished memory of her beloved Fouzia. The man is a gem dealer and he has a good look at my ring. He doesn't comment: he just sniffs.

Maha's stepfather is there too. He has his back to us. He's sulking and I can feel the anger radiating from him. They're talking about Nena—about what to do with her. Should she go to London or Dubai? Or should she stay in Lahore and hope the marriage with Mota will materialize? No one is sure and opinion is divided. Maha doesn't have an opinion on anything and I have to maneuver her back home so that she can collapse into her own bed. Maha's mother looks at her daughter slumped on my arm as we stagger out of the door. She shakes her head, turns away, and lights another cigarette.

I'm fed up with Maha and I can't tolerate her insistent belief in black magic. Nisha danced this evening. She put a massive effort into it, but it was a brittle, staccato performance. We were all watching: Maha, me, the children, Master Jee, and one of Maha's cousins. Nisha bent backward and we all clapped. Then she flung herself forward. We would have cheered except that she kept falling toward the floor until her forehead made contact with the rug and she collapsed with a piercing scream. The family crowded round and Maha cradled her head and kissed her.

"She's too hot," I said fetching some water from the fridge. The air cooler wasn't switched on and it was a close, humid evening.

"It's not the weather," Maha retorted. "It's black magic. It's her," she shouted, nodding her head in the direction of her mother's house. "She's jealous." Jealous, she means, because Nena is to be married and will bring money into the house.

I said I thought that Nisha collapsed because she's in poor health and because she became overheated and exhausted. Nobody thought it was a sensible explanation. I was corrected by Maha, Master Jee, and the cousin, who all affirmed that an evil spell was at work. Nisha corroborated the black magic story and supplemented it by claiming that a man in black had entered the room and had asked her to go with him.

I thought to myself that I'd like to leave with the man in black so that I could escape this idiocy. Nisha was crying, the cooler still wasn't switched on, and Maha was working herself into a panic, bewailing the many spells placed upon her family. I won't go back tomorrow; we'll only argue.

The Bride

I've decided to relent and go to see Maha. I've bought a big bag of *sita*, corn cobs, from the man by Roshnai Gate who cooks them in a pan of rough salt over a fire. He buries them in what looks like gray grit and then stirs the cobs around so they become crispy on the edges and encrusted with salt. I have a dozen of them sweating in a *shopper*. They smell delicious.

A red-eyed Nena throws the door open and begins to cry.

"Louise Auntie," she croaks, "I'm going to be married tonight."

I hug her and ask if she's happy, and she shakes her head. "I'm happy for my family."

There are some people in the best room and I ask if one is Mota, the lucky bridegroom.

"It's not Mota. I'm not marrying Mota. I'm marrying Sheikh Khasib. I'm going to the Gulf," she sobs.

I sit with Maha and Nena in the other room. Nena is crying and doesn't look as if she'll make a very happy bride.

"She's too young," I argue. "And it's too dangerous. Anything can happen to her."

Maha has twenty thousand rupees ($337) rolled into a tight tube in her hand. It's an advance on the one and a half *lakh* ($2,529) that Nena will receive for having sex.

"I'll give it back to the *dalal*," Maha promises.

"No, Mum. I want to go. I *do* want to go. Then we'll have some money."

Nothing has prepared me for the reality of this event. I don't know how far I should oppose the decision. I'm here to document life in Heera Mandi, not to intervene in it. In the social codes of Heera Mandi, Nena is not doing anything unusual. By the standards of the *mohalla* she's not a child: she's ready for "marriage" and she's ready to become a *kanjri*. I know this on an intellectual level, but it doesn't lessen my unease. I've become too deeply involved in the lives of the people I'm supposed to be researching; I'm no longer an objective observer but a participant in their world. I can't walk away from this situation without losing my integrity, but I can't stay and keep it either.

I could stop her going. I could create a scene, but even if I oppose her marriage in Dubai it will only postpone it for a short while. If I go to the police Nena won't be protected: she might even lose her virginity in the *thanna*. I could contact a charity, in which case she might be removed from the family she loves and placed in a government home or a shelter. And, even then, her future wouldn't be secure because marriage is the only future for a Pakistani woman of her class. And the homes themselves are unsanitary and draconian — more like prisons than refuges. I don't know what is best for Nena, or for Maha, nor do I know how I can protect the child without betraying her mother, my friend.

The people in the best room are the *dalals* — agents, or procurers. The principal *dalal* is a woman; she is with a man who is a kind of minder. She calls us and says we have to hurry up. We must leave the house in two hours and first we need to go shopping.

"Ring Laila. Ring Mota," I suggest. "Maybe she can still be married here."

Maha shakes her head. "Mota hasn't called for days. He's not interested."

The *dalal* takes my arm. She speaks impeccable English and tells me that she's Pakistani. She grew up in Dubai and still has family there. She wants me to help calm Maha. "I know the mother is worried," she says, "but there's nothing to be concerned about. She's only going for one man: for Sheikh Khasib. No one else will place a finger on her." Perhaps this is supposed to comfort us. This woman thinks I'm a *dalal* too. She believes that I take girls to England. She speaks about the trade as if we are business executives sharing notes.

"You know Sheikh Khasib, don't you?"

I say I know of him — he belongs to the inner circle of royal dynasties and is one of the most powerful men in the Gulf states.

The *dalal* fills in some the details. "He likes virgin girls. Not for sex—he has lots of girlfriends from all over the world for sex—but he likes to open virgins. It only takes a few minutes. The girls come from all over: from India, Pakistan, Iran. He doesn't even take off his clothes for most of them. It's like a habit for him. Someone told him that opening virgins makes a man young."

"Opening virgins" is a common way of describing defloration, and I've heard this stupid myth plenty of times before. It makes Nena sound like a vitamin tablet.

I ask her some questions about the travel arrangements.

"The passport, the visa, and her identity card have all been arranged." She hands the documents to me. They look real enough. "If there's a problem with them it'll be a problem for me. I have to think about my reputation and my business. This isn't the first time we have done this and it won't be the last. I've taken ten or more girls from Heera Mandi and it's easy. There's no problem. She'll be met in the Gulf by my sister and she'll stay with my family. She'll be sent to Sheikh Khasib tomorrow night, and afterward she'll come home."

This woman is part of a well-established network that supplies girls to rich Arab men. Her reputation depends on her ability to procure beautiful virgin teenagers. She knows what her clients like. "She has to have a nice dress, something that shows off her figure. He likes to see the shape of their bodies. And no embroidery and fancywork. He doesn't like the feel of it. Just something simple and very, very sexy."

Maha still has the money in her hand. Some of this advance is going to be spent on purchasing the sexy outfit. Maha agrees to be back at the house with the special dress in an hour and a half. The agents say that they'll return with the plane ticket and we'll all go to the airport.

I don't relish the prospect of helping to choose a dress for Sheikh Khasib to savor, so Maha and Nena go shopping without me. They return from Babar Market with lots of bags. Nena has a new leather suitcase for the journey. She has two pairs of fake gold-and-diamond earrings that would never pass for the real thing. She has lots of lit-tle sachets of shampoo and a pot of cream bleach for lightening facial hair. And she has a new outfit—one that the sheikh will not like. The *dalal's* instructions have fallen on deaf ears. The dress is baggy, a woman's dress, but Nena has the body of an adolescent. The dress isn't simple either: it's black and heavily embroidered with silver thread and silver and glass beading. The shoes are worse: three-inch platforms and six-inch heels covered in thick silver glitter. They look like a poor girl's party clothes. The agent sighs heavily and says that she'll arrange for a tailor to do something with the dress in Dubai. Maha and Nisha bring out all the clothes in the house in the hope that one item will be suitable for a palace. All are variations on the same theme: day dresses in cheap cotton or polyester party wear with embroidered panels and sparkly bits. A pile of items is shoved into the suitcase in the expectation that something will be right.

Nena has a new outfit for the journey. It's black cotton with a peach *dupatta* and embroidered flowers around the neckline. She looks so pretty. She's wearing eyeliner and a little lipstick. "Keep it light," the agent says. Nena doesn't need to attract attention by looking like a prostitute on the move.

"Will I be the only person on the plane?" Nena asks.

Nisha, Ariba, and Sofiya aren't going to the airport with us. Nisha is terribly agitated. She's moving round in the other room with strange twitching movements. The minder takes Nena's bag and the two girls stare at each other and then they hug. Nisha is crying and her thin arms are locked around Nena. She won't let her little sister go and the *dalal* has to pry them apart. We can hear Nisha shouting for Nena as we go down the stairs. She's hysterical. Ariba and Sofiya are waving goodbye to Nena from the balcony, and a crowd is assembling to see us leave in a taxi.

Nena sits between Maha and me in the back of the car. The *dalal* is in the front and the minder rides alongside on a motorbike, with Mutazar sitting on the gas tank. Nena is resting her head against her mother. She's not making a sound, but she's crying and her eyeliner is running down her cheeks. Maha wipes away the tears, one by one, and kisses her daughter's hair. Neither of them want this marriage to happen, but the money is good, and Nena's reputation and the family's honor will be enhanced if she is known as the virgin bride of Sheikh Khasib.

I'm confused about the direction we are taking to the airport. We stop in a poor neighborhood and the *dalal* tells us that Nena is going to the doctor's for a checkup.

We pass some women sitting on a wall. Their body language and direct eye contact make me think that they're prostitutes.

Maha is seething and muttering curses. "Such an insult. That sister-fucker woman."

I ask her what's going on, but she's too angry to reply.

The doctor's surgery is a small dark room in a large dark building. There's no evidence of its medical function except for an ancient heavily stained mattress and a plastic tub containing a few bottles of liquid, some gauze, and a couple of syringes. The surgery makes Dr. Qazi's practice look like cutting-edge medical science. The doctor is female, in her midthirties, and heavily pregnant. She says she needs to give Nena an internal examination to make sure everything is in good order.

Nena is terrified, but the *dalal* clucks around her and persuades her to lie down. She squeaks and cries, holding onto her mother's hand. The doctor tells the *dalal* that the sheikh is going to have difficulty because Nena is so *choti*—so little, so young—but that the girl is definitely a virgin.

"We have to do this," the *dalal* explains. "It's protection for us and for the girls. We don't want to take them all the way to Dubai and then have complaints that they're not virgins."

This doctor is a *dai*, a traditional midwife who specializes in treating women in the sex business. She also deals in abortions and in verifying virginity for fussy clients. She prepares a syringe, explaining that Nena has a problem and needs medicine.

I ask what the problem is.

"She has too much water on her uterus. She's been eating too much meat and hot foods."

It sounds bizarre and I ask Maha what it means.

"She's weak. There's too much water." She speaks to me as if

I can't understand the Urdu, when what I can't understand is the medical condition.

"Louise," she repeats. "Nena is weak."

"It's the Pakistani diet," the *dalal* adds.

I think this sounds like good news. Maybe she won't have to go if she's ill. The *dai* gives Nena an injection in her buttock. She uses some solution from one of the bottles in the plastic tub. Lots of other girls must have this water problem too. She tips some white powder onto a piece of paper and tells Nena to eat it. It can't taste nice because Nena is pulling a face. The *dai* pours more powder onto another piece of paper and folds it up. Nena has to take half tonight and the rest tomorrow morning. She's told that it will make her feel a lot better. The *dai* is right—Nena doesn't cry any more.

The airport is busy. Hundreds of people are clogging up the entrance and travelers are fighting to get through the doors with mountains of heavy luggage on carts. We can't go into the building without a ticket, so we stand by the refreshment stall. The *dalal* is tutoring Nena on what to say—the name of her husband and the reason for her trip to Dubai: She's 18 and she has been married for a couple of months. Her husband works in Dubai and has asked her to visit. She's told not to worry: it's going to be easy.

I can't believe I'm hearing this. I speak quietly to Nena. "You don't have to go. Please stay. It's possible." She smiles and says she's fine. She's calm and very dignified and Maha and I are the ones who are crying.

We watch through the windows as Nena checks in. She grips her suitcase and turns around to wave and blow kisses. We stand here for an hour, hanging over the barriers waiting to see if she gets through immigration. It's oppressively hot. And then the *dalals* tell us that everything is okay—she's on her way and we can go home.

Maha's house is unusually quiet. Maha is talking about their future life—about putting a deposit on a house with Nena's *"kusi* money." She pulls the roll of notes from her bra and spreads them on the mattress. Part of the advance has already been spent on Nena's new clothes. "This will pay for the rent," she says putting a pile of notes on one side. "And this is the electricity money."

Farrukh, the helper, has bought some special food from a restaurant on the main road. It's delicious: freshly barbecued chicken with green chilies and the thinnest *roti*. Maha pulls a face and waves a piece under the children's noses. "We're eating your sister's *kusi* money." And then she turns to me and adds, "In Heera Mandi your *kusi* is your gold. Nena has a golden *kusi*."

She stops eating. "Louise, were Nena's clothes all right? Was her bag all right? It was a nice bag, wasn't it? And the shampoo, was it good?" There's a note of panic in her voice. I nod and Maha crumples and cries. The others carry on eating in silence.

A dnan has arrived. I haven't seen him for months, and neither has Maha. He's been in the hospital. His legs blew up like balloons and they thought he was going to die.

He's drugged up and vacant, and there are three syringes sticking out of the pocket of his *kameez*. Maha isn't pleased to see him.

"You've been taking those injections, haven't you? While you've been sticking those injection in your ass, my daughter has had to go to Dubai."

There's disbelief on Adnan's face and he sits with his head in his hands. He's known Nena since she was 8 years old.

Maha is shouting at him. "We didn't have any money. No money for the rent. No money for food or electricity. You gave us nothing. When you're in hospital, or at home with your wife and your drugs, we still need to eat."

The phone rings and she screams into it. It's Mota, the reluctant bridegroom. "My daughter wanted to marry you, but you didn't come," she shrieks. "She loved you and now she's had to go to Dubai."

Adnan is swaying to and fro. He staggers to his feet, swearing at

Maha. "You *kanjri,* you *taxi,* you *ghashti.* You're a rotten mother for sending your little girl to Dubai."

"She didn't go because of me. She went because of you. You weren't here and we were hungry."

Adnan leaves the door wide open and staggers down the stairs. He says he'll never come back.

"Good," Maha shouts after him. "Go and die, sister-fucker."

m ota arrives at midnight with a friend. He doesn't believe that Nena has gone to the Gulf. Perhaps he thinks it's part of Maha's negotiating strategy. He rings the airport to see if a plane had been scheduled to leave for Dubai at eight o'clock. He lies on the mattress with his head resting on his hands. He isn't convinced.

He's wearing a rather nice Western outfit—a beige polo shirt and extralarge cream jeans that are fastened underneath his belly so that we have frequent and generous views of a squashy, hairy stomach. He refuses food and drink, and after reviewing the situation for half an hour he hauls himself up so he can sit on the edge of the mattress with his back to us. He's wondering how he's going to get up without groaning. He rocks slightly to get some momentum going and then he stands. As he does, his jeans slip down revealing giant buttocks. The jeans would have fallen to the floor, but they're caught on his penis. Mota wriggles back into the jeans with a little exclamation of surprise, and he and the friend leave.

Everyone explodes into laughter the moment they walk through the door. "What an ass," howls Maha.

"We saw Mota's ass," Sofiya gasps.

We're crying because the scene is so hilarious—the great, rich *ashik* exposed. Ariba performs a reconstruction of the event and the place falls apart.

"It was this big." Nisha spreads her arms wide.

"No, it was bigger than that."

"And it was really black."

"Mota's big black ass."

"So black. So big!"

I hope Mota can hear us as he levers his ass into his Land Cruiser.

Trafficked in Dubai

There's been no word from Nena all day. She was supposed to tele-phone: the *dalal* promised us. Maha is waiting for her call. I can see her from my window. She's staring into the courtyard, heavy-eyed and leaning on the air cooler that juts out onto the balcony.

I think about Nena and feel so guilty. When I write about prosti-tution and the trafficking of women in my office at the university, the issues seem clear, and yet here I am, in this dreadful place, wit-nessing a girl whom I am deeply fond of being trafficked to another country so she can be sold to a man who collects virgins. I've spo-ken to a doctor about the visit to the *dai*. He's never heard of the water-on-the-uterus condition and says it's nonsense. The injection was probably a muscle relaxant—something to help a child cope with sex with a man. And the powder was probably an opiate to keep her calm. I'm unable to analyze this and be objective: I've lost my pro-fessional moorings.

It's night in Heera Mandi and it will be night in Dubai. Nena will be going to her husband. Maha is lying on the floor clutching her stomach, and all the members of her family have gath-ered around. They think she might die.

"I have a pain in my belly," she groans. "It's like a delivery pain. It's so bad. Like cancer."

She doesn't die and, when the relatives have gone, she crawls to lie on the mattress with me. She wants to talk about Nena, but it's a mixed-up story and I struggle to follow. It's a story about a handsome young man who told a girl that she was beautiful and sexy and kept her in his gold bed for a month. It's about pain and blood all over the bed and about the drugs that make her remember only half of what happened when the man took her to the big golden bed shortly after she had married Sheikh Zayed, her first husband. It's a story about Sheikh Khasib who, more than twenty years ago, enjoyed a 12-year-old girl from Heera Mandi. That girl was Maha.

ena's marriage to the sheikh didn't happen. She telephoned us today and said the sheikh didn't want her. The *dalal* says something different. Nena was presented to her husband but was so ill with stomach pains that she vomited and had to be taken away. They'll try again tonight.

I've contacted a friend who is a leading Pakistani human rights lawyer, and he says we can launch a search for Nena. No one in Heera Mandi thinks this is a good idea. In fact they think it's an atrocious idea. They don't want the authorities involved. They say it'll only cause them more problems and they forbid me to take any action.

Twenty-four hours later, the *dalal* is angry and Maha is shouting back into the phone. The *dalal* says that Nena is refusing to cooperate. She is sick again. The girl is hysterical, and she has to stop the stupid behavior and earn some money. Until then she won't be allowed home. Maha replies with a stream of abuse.

Nena is distraught. She's screaming down the phone. I can hear her from the other side of the room. She hates her mother, she hates Dubai, and she never wants to come back to Lahore. I can hear her shouting, "You're a horrible mother. You want me to die."

Maha is shaken by the call but is trying to be positive. She says we'll have a party when Nena comes home. We'll have wonderful food and Nena will have a beautiful wedding dress because, after all, she will have been a bride.

Nisha fumes, "You never got me a wedding dress."

Maha and Nisha start to argue. Maha grabs Nisha and shakes

her. Nisha is screaming and garbling, "Give me back my *lakh* of rupees. Give me back my *lakh* of rupees."

I am startled: Nisha was married too, in the days when she was well—two years ago, when she was 14.

*A*nother two days have gone by and Maha's younger brother is giving instructions to Nena on the phone. He tells her to be calm. Once she's gone through with the marriage, she can come home. He won't let me talk to her. He says she'll only get more upset.

Maha has taken an overdose of sleeping tablets and can't talk. Nisha is speaking for her mother. She whispers that they went to a "deck function" last night. The clients had a smart bungalow with a swimming pool in Defence. Her uncle—the one who is speaking on the phone—acted as the agent. He'd been asked to organize the entertainment for the party. He rounded up women from the area: Maha, Nisha, the daughter of the Pathani *upar walli,* and three others. He told them it was an "open-price event." The women would have to work hard and earn money from the clients without any fee being fixed in advance. He brought Nisha a sexy dress to wear. It was sleeveless and backless: she must have looked shocking with her sticks of arms and every vertebrae in her back poking through her skin.

The men were drunk when the women arrived at midnight. There were lots of them and the function was out of control. They locked the doors and the women were stripped, kissed, touched—and more. Nisha can't say. When they left, Maha and Nisha were given one thousand rupees ($17) between them. It had been a fixed-price function after all and Maha's brother had pocketed the rest of the fee.

The Pir *and a Message for Nena*

A *pir* has been in Heera Mandi for a couple of days. He travels with a drummer and a man who has a very loud voice. They announce his arrival and ask for donations from the locals. He's taken up residence in the *dalal*'s hut and Maha and I are going to pay our respects. Farrukh, the servant, is sent to buy some sweets from the bakery and we join the holy man in the pimps' den.

The scruffy *pir* is sitting cross-legged on the *charpoy*. He's in his late fifties, with long hair in a confusion of untidy dreadlocks. A big silver *panja* hangs around his neck and he has lots of rings and a couple of massive ankle bracelets. He has kind, laughing eyes in a thin face and a big gold tooth set in among his other long yellow ones. He speaks nicely to Maha.

Farrukh brings in the sweets and the *pir* blesses them and hands them around. We sit for two hours drinking tea and eating the sweets. Maha explains her problems, and the *pir* listens patiently and responds with kindness. I don't know if he's holy, but he seems like a nice guy who has a lot of time on his hands and who is willing to talk to people about their problems and offer some sensible advice. He tells Maha to forget the black magic and to pray. He says that she should stop entertaining the *tamash been* because none of them are any good. "They use a woman: they pick a flower and they throw it away." Maha nods as if enlightened.

Mushtaq, the pimp, marches into the hut and we fall silent. He's radiating power and authority. He stands in front of a broken piece of mirror balanced on top of an old cupboard and combs his hair for five minutes. Then he marches out, taking his aura with him. The *pir* suppresses a smile: he doesn't like Mushtaq. Not many people do.

Maha hangs on the *pir*'s words. She reminds me three or four times that he's a *sayeed*, from Iran. He's so gentle and patient with her, so nonjudgmental. He knows she's a *kanjri*. There's a softness in his eyes as she tumbles over her words. She explains about Nena, her little daughter lost in Dubai, and Adnan and his injections and how her children's luck is turning bad, just like her own. He suggests that they should move. They should concentrate on keeping the house clean and living simply. He makes jokes so that Maha laughs in between her tears. I came into this hut prepared to be cynical, but I will leave impressed by a man who has given genuine comfort. He's a psychologist and a therapist and he has given Maha real hope.

He promises Maha that he'll help Nena: he'll pray and ask God to put love and gentleness into the soul of her *dalal*. He himself will also speak to the woman's heart so that she will soften and send Nena home. And because Maha wants to do something more, so that she can feel she's helping her child, we buy a pure white dove. The *pir* holds it in his hands and blesses it and Maha makes a prayer, and as the bird is released and flies high into the air, she blows it kisses to take along with her love and the *pir*'s blessing to Nena in Dubai.

Dancing Daughters

A thick fog hangs over Lahore and every
other person has a nasty rattling cough.
There's no crisp frost or biting wind, but it's
cold and there's a heavy dankness about
the old city. The houses never seem to dry
out—there's not enough sun or warmth to
drive away the damp. The chill seeps into
walls, clothes, and bones. The men drinking
tea on the benches outside the Al Faisal Hotel are wrapped in layers of shawls and
blankets. The old people complain about
their arthritis and the tuberculosis patients
hack. When I'm away from Heera Mandi,
I remember the intensity of the heat in
May and June and the paralyzing humidity
of July and August. I always forget the
cool, slightly depressing depths of the winter mists.

The view from my room is in soft focus.
Fog forms a white veil over the courtyard

and its buildings. Most of the shutters are closed, the old ladies haven't set up their charcoal burners, and the door to the pimps' den is shut, although a strip of blue fluorescent light shines from underneath it. Maha has moved. A new set of ragged sheets and clothing is hanging over her old balconies. The plants have gone and so has the noise from her tape deck.

Dusk is falling and the rooftops of old Lahore are being lit. The lights on the top of the *panje* are switched on one by one. Among the trees by the *masjid* the addicts crouch around half a dozen fires, and by Roshnai Gate, the snack *wallas* are preparing new warming treats to tempt us: roasted peanuts and sweet potatoes baked until their skins turn black. By eleven there is an unusual calm in the courtyard. The ice-cream man doesn't come around, the popcorn *walla* stays for no more than a few minutes, and the only sounds at midnight are those of the odd bike or rickshaw and the clicking of the *malish karne walla's* bottles of massage oil on their metal tray.

War Reaches Heera Mandi

This morning the Hazoori Gardens are almost empty. A couple of dozen men are lying on the ground covered in blankets. There are only a handful of cold-drink sellers and no one is offering to take my picture today. The mist is beginning to clear, and although the sun's rays are feeble, they are enough to lift the spirits. The entrance to the fort is prettily pink, washed with watery sunshine. The trees with their few remaining leaves are beginning to look picturesque rather than barren. Unfortunately for the tour guides and the drink sellers, there are few tourists here to see the fort in the wintry sun. No one wants to come to Pakistan these days.

Pakistan has had very bad press in the West; television and newspaper reports back in England seem to equate visiting Pakistan with a death sentence. It's portrayed as a hostile battleground prone to horrifying, bloody violence. You'd imagine the place to be full of religious fanatics and committed terrorists. This is not the Pakistan I know.

Despite my positive feelings about Pakistan, it's a very bad time for Americans or Britons to visit. At home, everyone says I shouldn't come here. The West is waging a war in Afghanistan and a minority of people in Pakistan hate the Americans and the British even more than before. Although the most extreme of the *mullahs* are under arrest and their supporters are lying low, I still feel apprehensive. I'm frightened of big men in Pathani-style turbans, and even more scared of small men in Pathani-style turbans. I have an unfounded but hard-to-shake feeling that small, thin ones are the angriest.

Things are tenser in the north and along the border with Afghanistan, especially in Peshawar and Quetta. Many Afghans have settled here, and some Pakistanis have close family ties and economic and cultural bonds with the Pathans, who are the Afghan ethnic group leading and supporting the Taliban. Here in Lahore, though, India is the main enemy, not the United States and Britain. In Heera Mandi, the United States is considered the land of plenty and, despite the war, I am still asked for help in obtaining visas to London, the next best place.

Few women in the *mohalla* have even the vaguest idea about world events, and they know virtually nothing about their own country. They don't read newspapers and their radios and televisions are tuned to music channels. They do hear things through the grapevine. A couple of women have heard that there was an enormous fire in the United States and big buildings collapsed. The men know a lot more: they insist that Muslims were not responsible and that Osama bin Laden had nothing to do with the atrocity. Muslims, they say, are disorganized and technologically incompetent. They claim September 11 was too grand and too organized to be the work of Muslims: it had to be the work of the Jews.

My sense of personal danger has been lessened a little by a new episode in the long-running Indo-Pakistani quarrel. There's a round of international diplomacy and troops are massing on the border. I'm confident that it's nothing more than the usual posturing made a lot sharper by shifts in the subcontinental power balance. Thanks to the war on terror and India and Pakistan's testing of nuclear devices in 1998, Pakistan is important again in American foreign policy and world politics.

The Indo-Pakistan tension is rippling though Heera Mandi in a disturbing way: all the Indian television channels have been blocked. Among the things likely to motivate the women of Heera Mandi to action are religious events and, now, the absence of Indian television. Those much-loved soaps and twenty-four-hour Hindi pop programs are off the air, and no one has yet managed to retune their sets to other exciting channels. Instead we have to watch Cartoon Network and the BBC. On Pakistan Television (PTV) we watch an eminent chemist talking about the difficulties and joys of fish farming. Maha looks askance and asks, "Do you have to watch this rubbish in London?" We are desperate to get the pretty girls, lovable macho men, and spangled dancers back on the screens again.

In Tibbi Gali, Shela is in her shop crouched over a charcoal burner. The young, black-skinned girl who looked so nervous last year has changed. She's eager to talk, laughing, and asking where I've been and if I can bring her some shampoo when I return. Shela wants to know how much things cost in London and America. She's shouting, "How many dollars is shampoo in America?" I'm cringing. A big group of men is congregating around me, and I really don't want to be identified as American in the middle of Tibbi Gali when a war is being waged in neighboring Afghanistan. She shoos the men away and gives me tea. She says business is down these days, and I ask why.

"It's because of the war," she explains and I'm confused. "Don't you know anything?" Shela says in surprise. "There's a big war between India and Pakistan. All the men have gone to fight."

A New Home: A New Beginning?

Maha's new house is a fifteen-minute walk away—just on the edge of the old city in Karim Park. Maha thinks they are making a fresh start and that maybe their luck will change. It's a respectable *mohalla* and the house is on a cleaner road that has only a few open drains and potholes. They have two rooms, a bathroom, and yet another small, waterlogged kitchen. It's basic but a lot more sanitary than their last house. Best of all it has an open yard, rather like a well in the building that offers light and fresh air. Maha's home is on the ground floor, so she has appropriated the area as her own. The concrete floor has been swept and she has put her plants against the wall.

Maha is dressed up for my visit. Nena is walking around in a very old, extralarge-size man's jacket. She looks the way she always does—pretty and a little ill. Nisha seems pleased to see me but then returns to lying down, morose and thin, on an ancient sofa cushion in front of the gas heater. Ariba is neater than I remember, and Mutazar and Sofiya are jumping around deliriously because they know I've bought chocolates from England.

One of the rooms has been carefully prepared. Maha has made big cream satin cushions and a bedcover. She has bought a new rug, which no one is allowed to stand on, and a vase full of plastic flowers. We lie on the new bedcover. Maha says some ladies from Heera Mandi wanted to rent the room so that they could entertain customers, but she said no: it's her special room and she doesn't want anything to spoil it. This is a good *mohalla*. A delegation of local residents has already been around to ask her to turn down the music at night. It doesn't set the right tone, and they want the *mohalla* to stay respectable.

Nena is lying next to us eating *namkeen*. She returned from Dubai after six weeks. In the end she didn't marry Sheikh Khasib. Every time she was about to be presented she became hysterical, so she was sent back to the *dalal* in disgrace. She met ten other girls who had also been sent for the sheikh. Most were 15 or 16, and she says that they were all really beautiful. Nena got to know three others well: a Turkish girl, an Iranian, and one she described as an Arab. It was a first marriage for all of them.

At least the *dalals* didn't force Nena to oblige other clients in order to recoup their money, but they didn't treat her well. She had to dance in a club every night for six weeks to earn her passage back home.

"She was so thin and she had three hundred lice in her hair," Maha cries with indignation.

"And Ariba has been trouble," Maha adds. "She was raped again and this time she got pregnant. She had an abortion. It was terrible."

The abortion was six weeks ago. Now Ariba is going into the business. Maha reasons that she might as well get paid: when they rape her, the men get Ariba for free and the girl deserves more respect than that.

Mutazar is having fun devising tortures to inflict on a tiny, bright yellow chick. I ask him not to do it and I've threatened him with retribution, but it hasn't made any difference. He takes pleasure in wrapping it up in newspaper, putting it in a cup, dangling it by one leg, hitting it with a variety of implements, dropping it, and throwing it in the air. When he stops for a second it stands with its eyes closed, its beak open, and its entire body heaving. I wonder how long it will take to die and hope it will be soon. I'm amazed that it has lasted this long. It must be robust and have exceptionally pliable bones.

"Agh, Mutazar is so happy," Maha says pulling him toward her and kissing him. "He's got a friend to play with. Tomorrow we'll get you another chick."

Maha has a new helper whom they call Baaba (father). He's an elderly laborer working on the house and he runs errands in return for a meal and a chance to sit by the warm heater. Baaba has tattered clothes and the look of the apologetic hungry. He's like Farrukh and Ama-Jee before him: poor people who eke out a bare existence by being informal servants and errand runners for the slightly better off.

Maha gives him a compassionate smile and slides an extra piece of potato onto his plate.

Nena is giving us a dancing show. She's doing less of the classical repertoire and more of the Hindi pop routines. To the modern man's eye it must seem more exciting. She'll make a lot more money this way than by an hour of stamping with her *ghungaroo* and doing pretty things with her hands. She shakes a lot, swings her hair around, and has perfected the cleavage-revealing maneuver. I'm disappointed, but I don't have the heart to say so.

Maha glances at me constantly to gauge my reaction. She's so proud of Nena, but after the dancing ends a miserable cloud settles on her. "I'm finished, Louise," she says. "Men don't want me. They want my daughters. My time is over and now it's their time." She hasn't reconciled herself to the idea. She misses the power she had as a beautiful young woman. There's only one thing she can do now that her power is waning: she'll become a *naika* and deal in her daughters.

A *Modern* Mujra

The chick died last night. Maha says it had a pencil rammed down its throat. Mutazar smiles and hides his eyes behind his hand and mumbles in a baby voice, "Ariba did it."

Mushtaq the pimp had better watch out. When Mutazar leaves the women's world to join the public world of men in a few years time, there will be another long-indulged bully lounging on the *charpoys* outside the pimps' den.

Nisha and Ariba are not at home. A *dalal* collected them in a rickshaw because a customer wanted to inspect them. If he approves of them, the *dalal* will make arrangements for a future liaison. Tonight is New Year's Eve and not an especially important night for most Pakistanis because it's a new year only according to the Christian calendar. Among some of the rich, though, it's a good excuse for a party, and the ladies of Heera Mandi will be in demand at a number of these.

I hope that Ariba and Nisha won't be the captive entertainment at one of Lahore's whiskey-fueled parties where the only guests are men.

Nena says Ariba was keen to meet the prospective *tamash been*. She had a wash and applied a dramatic amount of makeup. Nisha wasn't as happy. She sulked and refused to get ready. Maha shrugs and sighs. "Nisha hates men."

Sofiya won't leave me alone, kissing me and hugging me so that Maha shouts at her for bothering me all the time. "Keep away from Louise. Don't keep touching her and kissing her. You smell of wee and she doesn't want to smell wee."

We are having some fried fish as a New Year's treat when we are interrupted by a call from Laila, the well-dressed promoter and "bitch-woman." She's holding a little gathering and would like Nena to dance for her guests. Maha and Nena respond immediately. We have to be there within the hour and it takes forty minutes in the rickshaw. The meal is abandoned and the plates are left on the floor. Nena is told to put on makeup—lots and lots of it—and she's got to do it quickly.

It's a botched job and Maha shrieks, "What horrible makeup. You look like an old woman." She grabs a tub of cold cream, slaps a thick coating over Nena's face, and rubs it off violently with a towel. Then she does a proper job herself. Nena looks lovely, and Maha has even managed to erase the dark shadows beneath her eyes with some dabs of foundation.

We take a rickshaw to Garden Town. There are five of us in the back: Maha, Nena, and me, with Sofiya and Mutazar asleep on top of us. We are going to look very crumpled when we get out. Laila's house is modern and expensive, on a road with other nice modern houses. It's surrounded by walls, gates, a garden, and a spotless driveway in which two luxury cars are parked. It doesn't look quite as impressive inside. The big open hall has very little furniture: only a couple of sofas and, rather incongruously, a large chest freezer. The effect isn't so much minimalist as Spartan. Laila is plying her guests with drinks, and a lot of shouting is coming from the lounge. The party is small and select—just Laila and her husband, three other men, and a couple of women who are clearly in the business.

One is very pretty and knows it. The other is a little older, much fatter and plainer — and she knows it too.

The room has a half-finished feel. Painted in white, it has badly hung curtains, a blue nylon shag carpet, a glass dining table, and velour sofas with crude carvings on their wooden bases. Two small shiny pictures of Devon villages in white plastic frames hang on the wall, along with a clock in the shape of a big gold padlock.

Mota, the fat *ashik*, reclines on one of the sofas with the pretty woman. He's expanded since we saw his *bund* escaping from his trousers. He's wearing a safari suit and the jacket barely fastens around his middle. Nena, Maha, and I stand in the doorway feeling awkward. We're asked to sit on chairs by the table and are given drinks while Nena fidgets and tries not to look in Mota's direction.

The atmosphere is strained and artificially jolly: the party is supposed to be modern, more Western, and in a break from tradition, the male and female guests are obliged to interact. It isn't working; the men don't know how to speak to the women. Laila, wearing a short black tunic with tight black trousers and a diamanté choker of vast proportions, is trying to galvanize the guests by being extra loud and lively. She doesn't speak — she only shrieks and laughs. And she never walks across the room — she dances while making kissy faces at her husband. Something is unsettling her; she isn't the sophisticated, elegant woman I met a few months ago. She has too many big teeth set in a large painted mouth and keeps coming over to Maha and me and demanding that we dance with her guests, neighing at us like a horse. I say we've come to chaperone Nena. We're not part of the show.

Hurrah. It's midnight. Laila brings in a cake, shouting at me, "Well, say something. Say something."

I say, "Happy New Year," and the cake is cut. Laila's husband, a middle-aged man with tinted hair, a red shirt, and a floral tie, carries around a piece of cake and we all have to take a bite. He pushes the cake into Maha's mouth and into mine saying, "Here you are, Auntie."

Maha is fuming. "Did you hear that? The sister-fucker called us 'Auntie.' He's older than us."

Nena and the other girls are mobilized so that the men have someone to dance with. Mota is enjoying himself, waving his hands in the air as his stomach bumps against his partner. The two other guests are doing an embarrassed shuffle while staring blankly at the women.

Mota collapses back onto his sofa. He's drunk. He knocks over his glass of whiskey and the pretty girl rushes to refill it. It's time for Nena to dance and the floor is cleared. She starts her pop routine. She is very professional and executes some difficult steps, leaning backward so that her hair swirls into a shiny, blue-black pool on the floor behind her. She stops between songs to regain her breath, her mother and I dabbing the beads of sweat from her face and neck without disturbing her makeup.

The *tamash been* pay Nena by throwing money over her or by putting notes on someone else's head, or near their faces, so that Nena has to dance over to them and take the money and then drop it onto the floor with the other notes. In the intervals she collects the money from the floor and gives it to her mother who stuffs it into her handbag. We are doing the accounting, Maha reminds me. Mota is the most generous member of the audience: he's waving five-hundred-rupee ($8) notes, but he's so drunk he very nearly falls over the coffee table and into Laila's husband's lap. He places a note on another man's head and knocks over a lamp.

He's not much better when he is sitting down. He's overwhelmed with excitement, pawing the attractive girl and shouting, semicoherently, that Nena is lovely and a wonderful dancer. She's performed most of her routine with her back to him and when she bends over and wiggles her bottom one final time he lets out a giant bellow: "Nena. Nena." Even Laila is taken aback.

The guests are invited to dance again and Mota heaves himself out of the sofa and launches into serious foreplay with the gorgeous girl. Laila is doing an energetic dance that looks as if it's come straight off a raunchy music video. Then the pretty girl disappears and returns minutes later. She's changed into a tight black velvet creation with cutaway sides. She's not wearing a bra, and from the look on Mota's face and the way he's fumbling with her dress, he wants to find out if she's wearing any panties.

We get up to go. The music is switched off. Laila says that they would love it if we could stay the night. I'm sure they would, but we scoop up Mutazar and Sofiya, who are sleeping on the sofas in the hall, and flee for the rickshaw. The rickshaw *walla* is standing in the fog looking surprisingly perky. He's convinced that I'm in the business.

"You've been dancing," he says to me in a husky voice.

"No," I say.

"You've been dancing," he repeats with glee. He kisses my hand a dozen times and it's wet with saliva.

By half past two we're back in Karim Park. Ariba and Nisha aren't at home and they haven't called. Maha counts out the cash. Five thousand rupees ($84): it could have been better. "It was that bitch, Laila," Maha complains. "She told the *tamash been* to give one-hundred-rupee notes, not five-hundred-rupee notes." Thankfully, the drunken Mota hadn't heard.

Ariba and Nisha return next evening. Maha had to scour Heera Mandi for them. She found them in a *dalal*'s house. Nisha was presented to two men last night for an appointment—not for a prepurchase viewing. The *dalal* lied. Nisha says that the *tamash been* had a Land Cruiser: she could mean any kind of expensive four-wheel-drive vehicle. It had a fantastic little television in the front and the men had bottles of Black Label whiskey that they kept sloshing around. Another girl sat in the backseat with Nisha. She was very pretty and was wearing "pants shirt"—Western trousers and a top. She was nice to Nisha but the men were not so friendly. They said Nisha was too thin—they liked plump girls—so they took her back to the *dalal*. Nisha is lying on the mattress and laughing hysterically. She's talking about a sandwich and Maha is saying something about there not being enough meat.

Ariba went to the Pearl Continental. God knows what the *dalal* was thinking. Thirteen-year-old girls with lice, bitten fingernails, and rudimentary manners are not welcome at the Pearl Continental. Not officially. She had to stay there all night because there were police roadblocks to check for drinking and womanizing and it was dangerous to be on the move on New Year's Eve in 2001. We all could have ended up in the *thanna* short of several thousand rupees.

Nisha and Nena are sticking decorations in my hair and Maha is semiconscious. She's been taking Eighty-One again. Mutazar is kicking her in the back and Sofiya is chewing a pen refill. Her mouth is full of black ink and she thinks it's funny to spit the ink onto the bed so that it sprays into big splodges on the freshly washed sheets. She wipes it in fat lines over the fridge standing in the center of the room. Maha mutters something about the mess, but she's growing quieter by the second. Ten minutes later she notices nothing except the colors of the flames on the front of the portable gas heater.

m aha is on the verge of tears again this evening. The girls are fighting and Maha has no reserves of patience left to deal with the noise. "Go and die, bitches," she shrieks. "I swear, Louise, they will be the death of me."

The phone rings and Maha is all sweetness. She puts down the phone and snarls, "Sister-fucker." It was a big *dalal* who needed two girls but was offering only two thousand rupees ($34) for the entire night—for both of them. This is an insult. Nisha and Ariba are only 16 and 13: they're in their prime earning years but they aren't prime material: Nisha is a bag of sharp bones and Ariba looks like she's been pulled off the streets. Maha has refused the offer because she needs to maintain her girls' price. They're not that desperate yet, but the *dalals* know Maha's situation and they drive a hard bargain.

Maha is thinking of other ways to raise cash. She's set up a dress-designing service and is going to run up expensive creations to sell to boutiques for the rich. She practiced sewing when she made the new bedcovers, but the dresses are more technically demanding and she looks wearily at a billowing, wrinkle-hemmed, polyester *kameez*. I know she has considered going back into the business herself, but it would be unthinkable. She'd have to travel out of Heera Mandi so that no one here could witness her humiliation. She'd be no different than a common street prostitute, selling herself for a couple of hundred rupees, soliciting in the market, or waiting in a rickshaw at well-known pickup places for clients to climb in beside her. Her pride wouldn't allow it: it's beneath her dignity.

Nisha is sitting on the sofa, trembling and lifting her long, twisted limb. "No one likes me. I have a horrible arm." I don't know whether she's truly sad about this, but I think even Nisha, the man hater, is beginning to see that hope is running out for their family. No one is making any money.

Iqbal

Iqbal is worried too. I try to jolly him along, talking about all he has achieved and all the new challenges and exciting things ahead of him, but I'm making no progress. He doesn't want to sit with me on the roof terrace. He doesn't want to go out. Most worryingly, he's not painting and his easels lie empty. If Iqbal doesn't paint he will be lost: perhaps he will be lost even if he does paint, but at least he will be busy. He has risen above the ghetto of Heera Mandi without ever forgetting where he was born and the people with whom he was raised, but he is so scarred by this place and this life that I wonder whether he will ever find peace.

Emotional bonds are difficult for Iqbal to sustain because he wants them so very much, and yet, for all his life, the only bonds that he has known outside his immediate family are relationships based on sex and money. Nothing lasts in Heera Mandi, and love is a transaction conducted on the basis of an illusion. Iqbal knows this all too well, but he also wants more. He longs to love but he doesn't know how to, and he doesn't dare to learn—so instead, he paints. He paints the women of Heera Mandi because it is the only way he can truly connect with them. I wonder if he will forgive me for writing these words. I think he will because he knows them so well.

Ariba: Dancing Girl

In the middle of the afternoon Maha's house is still in darkness. Everyone is asleep. A big pile of dirty laundry sits in the corner of the room, and the pans, dishes, and leftover food have been stacking up for a couple of days. Nisha croaks that her mother went out alone yesterday and took a whole packet of sleeping tablets when she returned. Maha opens one eye to look at me and stirs slowly in the bed.

Ariba wakes up and limps around the room. She stepped on a nail in the old house when I was here in August, and the wound has become seriously infected. It's swollen and pus is oozing from between her toes. She holds her foot a few inches from my face and I back away.

Maha wants to talk about Adnan, not about Ariba's foot. I thought he'd gone for good, but he's still a lingering presence in Maha's life, bound to her because he needs a refuge from his official family and because he remembers happier days when he was still a vigorous lover. Their relationship follows the same old pattern, and Adnan has not been to visit or given Maha any money for three weeks. I get under the quilt with her and we work out her household accounts. It's a dismal picture. She doesn't have a clue about budgeting: she has a colossal monthly deficit just on her basic living expenses, and there are loans to pay off as well. She bought a selection of electrical goods from one of the traveling salesmen and has to pay him a thousand rupees ($17) every month. It's not too bad, she insists: he's a nice man and he might take payment in kind.

She has no jewelry left—she has pawned everything, even her earrings and the long gold chain she really loved. It was worth about thirty thousand rupees ($506), but she pawned it in the gold market for fourteen thousand ($236). If she wants it back she'll have to repay the loan plus another two thousand rupees ($34) for every month the broker keeps it. It's as good as lost.

Ariba has been lost too. She disappeared a couple of hours ago wrapped in an enormous shawl and hobbling slightly.

S he can't go. Look, it's her period." Maha is waving Ariba's blood-stained *shalwaar* at the *dalal* who lives upstairs.

Ariba is not having her period. It's just an excuse and the *dalal* knows it. The *dalal* isn't leaving—she has another strategy. She's cultivating a relationship with Ariba and trying to cut Maha out of the deal. We can see the *dalal* speaking quietly to Ariba in the court-yard. She tells her there's a rickshaw outside that will take her to Heera Mandi, where she can make a bit of money.

"We'll have to move again," Maha groans, "or else that *dalal* will eat my daughter." Ariba would have a new manager and the family would lose a source of income. Ariba limps back into the room and flops onto the floor. Maha is glaring at her.

She finally went to the doctor this morning and had her foot treated. It's been bandaged and she has to go for fresh dressings every day until the infection clears up.

"Go and do the dishes and clean the floor," Maha orders the girls. No one moves: Nisha is snoozing. Nena is watching a film and copying the actresses' mannerisms. "Ariba, *gashti,*" Maha screams.

"But my foot. I'm ill," Ariba cries. She pulls a face and rubs her hands on her temples. Her bandage will get wet: she'll be standing in an inch of water.

W e are taking Nisha to an appointment at the Zakariya Hospital. Her good hand is beginning to twist like a claw and she's complaining of stomach cramps. She can't even find the energy to dance with Nena.

The doctor is vague. He tells us that he is going to call the "bone doctor" and that we have to come back tomorrow.

"Is it serious?" Maha asks.

The doctor gives her a sideways look. "We'll see," he says and then adds, "The specialist is good: he's treated British people."

We've been in this hospital for hours waiting to see the consultant who treats British people. We sit in a cubicle to have Ariba's foot bandaged while we wait. There's a tiny cart, with bandages, cotton wool, gauze, a bowl of iodine, some scissors, and a pair of tweezers. They're dirty. A male nurse, who has just finished dealing with an old woman's leg abscess, begins to cut the dead skin away from Ariba's foot, and she's laughing and wriggling because it tickles. The skin is flying everywhere and lying in little wet, white curls on the floor with some crusts from the old lady's leg.

When we do see the doctor it's a short consultation. He says Nisha has juvenile arthritis. He writes out a prescription and tells her to sit in the sunshine for two hours a day.

"Will she get better?" I ask him.

"*Inshallah,*" he says. God willing.

The pharmacy gives us a bag of tablets. Most are vitamin tablets, and there are some expensive drugs too. The price of the cocktail is five hundred rupees ($8) for a week's supply. The consultant says she will need to be treated for three months. It's expensive, and I hope Maha will be committed to the regime.

We're walking back home, dodging rickshaws and motorbikes and getting stuck between *tangas* in the middle of the road. It's already dusk, and I'm giving Nisha a lecture on the importance of taking her medicine and doing what the doctor ordered.

"But, Louise Auntie," she complains in a slightly shocked voice. "I can't sit in the sun for two hours a day. Mum says it will make me ugly."

I t's not enough. Fifteen hundred [$25] is not enough for a full night. Ariba is *choti*. Three thousand minimum [$51]." Maha is sitting in front of the gas fire and haggling with a *dalal*. The two of them could be in the market trading carpets. The business negotiations are interspersed with tea, chit-chat, and information on a police raid that took place a couple of nights ago. The *dalal* comes from Heera Mandi and the two of them joke that this area is now "Little Heera Mandi" because so many *kanjri* have moved in.

Ariba sits next to us looking sullen and playing with a hairband. Her eyebrows are knitted as if she's bracing herself for a smack. She flinches every now and again even though no one is near her.

"What do you think, Louise?" Maha asks. "This is Heera Mandi life. A *kanjri*'s life." She doesn't expect me to reply.

"I can go," Maha suggests to the *dalal*.

"He wants a young girl."

"See, Louise," Maha turns to me. "See what the *tamash been* want." She's angry and is using a matchstick to clean her teeth with such ferocity that she's going to make her gums bleed.

Maha says he wants to try a child. He wants to "taste" her. She tells the *dalal* that she was in the business too when she was 12 and that her first husband was Sheikh Zayed. The way she describes it makes her sound like an old soldier talking nostalgically about heroism in battle.

"Three thousand," she says again.

The *dalal* shakes her head. "One thousand five hundred. That's it."

"Do you want to go?" Maha asks Ariba.

Ariba pulls a face. She's unhappy, but she doesn't want to say no. No one has worked since New Year's Eve, two weeks ago.

I touch Ariba's arm, but she shakes me off. Her mother tries to squeeze her, but Ariba pushes her away and nods to the *dalal*.

We're all sitting around the gas fire watching the changing colors. Pus is seeping through the bandage on Ariba's foot. Nobody is speaking. It's awful.

"Has Ariba done this many times before?" I ask Maha when we are cleaning the dishes.

"Three or four times. But this money is terrible. They say Ariba has a bad character. The men in the bazaar have looked at her—that's why her price is low. Nena's price is high because she hasn't done anything. She is *sharif*. A young girl should fetch five, six, seven thousand, but they say Ariba is *gandi*."

"What do I do, Louise?" Maha stops scrubbing the pans and looks at me. "The *tamash been* don't want me, and my big girl has a rotten arm. Look at it. And Nena can't go because it would spoil her and then she wouldn't be able to find a rich husband. When Nena gets married we will all be okay. Everything will be okay."

Nena's prized virginity may still be the family's temporary salvation.

"Ariba is young, I know it, but what can I do?" Maha sighs. "I say bad things to her, but I know she has a good heart."

Ariba is being prepared. Nena and Nisha have been sent out to buy a pot of cream bleach from the shop. It's for lightening facial hair but if you leave it on long enough it's supposed to make your skin whiter too. Maha mixes it on a plate and smears it all over Ariba's face and neck. She will have to use a lot because, as everyone so frequently points out, Ariba is almost black.

"Don't go," I say. "This man is bad—maybe worse than the other *tamash been*."

"No, Louise Auntie, "Ariba mumbles, "we need the money." She winces because the bleach is hurting.

I scrape the bubbling bleach off her face with a spatula and we wash off the rest in the courtyard. The others say she looks better. I can't see what they mean—she looks the same to me. She wants me to brush her hair and she gives me a hug. I ask her to promise that she won't go. I'm returning to Iqbal's soon to collect my things, and then I'll go to the airport for the flight back to England. I say that

I'll send some money. She promises she'll stay in the house and laughs a little breathlessly.

She rummages through the cupboard and pulls out one of Nena's old suits.

"Swear on the Quran," I say, but she only laughs some more.

I'm leaving. Maha waves from under a blanket: the Eighty-One tablets are just kicking in. Nisha and Nena are fussing around me and Sofiya is holding onto my legs. Ariba plugs in the iron and spreads the suit on a towel in the corner of the room. She smiles and turns away. She has work to do.

Old Ways: New Fortunes

(Monsoon: July 2003)

I haven't visited for over a year and Heera Mandi has been transformed during my absence. The local government has set about turning Lahore into a tourist destination: one of the city's bazaars has already been developed into a thriving, modern-day "food street" with open-air restaurants and quaint, upgraded facades on the traditional buildings. Something similar was intended for this part of the inner city until the guardians of morality objected, insisting that the tourists would gravitate to the brothels after satisfaction at the restaurant tables. Heera Mandi's women are disappointed by the abandonment of the plans, but now at least parts of the *mohalla* near the Badshahi Masjid look more like a visitor attraction than a shabby ghetto.

Fort Road has been paved. A fountain now sparkles near where the dump once stood, and there are decorative street lamps along the pavements where the addicts used to make their camp. Boys no longer play cricket in front of the mosque, and to make sure they don't return, two heavily moustached guards in dark sunglasses sit on a bench in the middle of the grass. The dingy facades of the buildings have been masked by layers of new paint: soft yellows and creams, dark reds, and occasionally a virulent, sense-jarring blue. The windows, balconies, and shutters have been painted in contrasting colors. At dusk Fort Road is picturesque, veiled in a flattering glow from pretty lighting, and the mosque, the Sikh palace, and the Hazoori Gardens are illuminated by hundreds of spotlights. It's utterly charming. On the side streets nothing much has changed: the potholes and filth remain and the *galis* are flanked by the same drab walls. The addicts haven't disappeared, but have simply gravitated from the perimeter of the mosque farther into Heera Mandi. A large multistorey building near the Tarranum Cinema has collapsed and now provides the addicts and other local men with an open-air toilet to replace the one that vanished when the dump was shifted.

All the buildings around the courtyard have had a facelift, and some of the people at the windows and doors are different. The village family has said goodbye to the courtyard. All the fancy furniture has been moved out. They have gone somewhere better, to someplace where the giant bed and dressing table can be fully appreciated. The family's fortunes have risen a very long way in a very short time. Five years ago they were in their village. Four years ago they were a hair's breadth away from destitution, living in two rooms opposite Iqbal's house and ridiculed by the Kanjar neighbors. And then it all changed. The girls went to Dubai, so now they can leave the bazaar in style, taking their new video player and three rickshaws full of electrical goods to a nicer home where the free-range chickens will have more room to roost on the furniture and scratch around on the carpet.

The Shia shrine in the corner of the courtyard has developed markedly during the past few years. It looks as if it has always been there. A metal fence has been added to the bathroom tiles, a gate has appeared, a couple of bushes are growing nearby and there are lots of places to stand little oil lamps. They look so pretty at night. The pimp who is the patron of the shrine must be very pleased. His religious life and his reputation in the community are coming along nicely. So is his business: there are three striking new girls in his house.

Maha's New Regime

Behind the freshly painted doors the interior world of Heera Mandi is undisturbed. Maha has a new home in one of the dressed-up buildings. She came back—again—to Heera Mandi after the police dragged her out of her home in Karim Park and kicked her down the street, beating her with a leather belt in front of all the neighbors. The police must have been short of money and anxious to generate some revenue through a cleanup campaign. Ariba and Nisha had a whipping too. Adnan rescued them from imprisonment in the *thanna* by paying a large bribe. He also paid the medical bills for the repairs to Maha's face. His money couldn't erase the damage to Maha's pride, though, and the family left Karim Park quietly one morning and returned to the stigma of life in the official brothel quarter.

Maha's new residence is far grander and cleaner than any of her previous homes. It's on the second, third, and fourth floors of a substantial building on Fort Road. A deeply devout Shia family live on the second floor of the house. The sons speak good English and have come to thank me personally for the overthrow of Sadaam Hussein, who was persecuting their Shia brothers. A less friendly family lives on the ground floor, but their door is always closed, so we rarely see them.

With three roof terraces Maha has plenty of space to spread out her possessions: her flowers, plants, and ancient furniture, the old air cooler, and the children's tricycles. It's a lovely collection of terraces: one is enclosed by trellises to screen the family from the eyes of the neighbors, and the topmost storey is so high it looks out over

all the other houses to Iqbal's restaurant, and the domes of the *masjid*.
A large metal *panja* and a Shia flag have been erected in the corner of
the roof. Below them, half a dozen oil lamps sit on a little ledge surrounded by strings of dried flowers and trails and pools of hardened
candle wax.

The seven rooms of the new house are strangely empty. Only
Maha and the youngest children are here. Nisha, Nena, and Ariba
have gone to the Gulf to work as dancers. Nena has been on several
tours and this is Ariba's second trip. Nisha wasn't so keen to go, but
the money was too good for them to refuse, so, still with her twisted
arm and bony body, she packed her suitcase and left with her sisters, all the while insisting that she wasn't going to have relations
with "old Hindus" or "black Arabs."

The advances on the girls' earnings are paying for the superior
house, and the family's fortunes have been revolutionized. They have
money and a lot of new furniture: odd multicolored table lamps and a
couple of giant beds with fancy carved frames that Nena's enamoured
ashiks have donated to the family in an attempt to win her virginity.
Maha deemed the gifts insufficiently generous, so two years after her
failed marriage to Sheikh Khasib, Nena remains complete.

I've left the beautiful room in Iqbal's home and moved in with
Maha. It makes sense because I spend so much of my time with her
and the family and my room in Iqbal's is often empty. I sleep on one
of Nena's big beds. Maha has assembled a collection of items on the
bedside cabinet that she thinks I'll need: tissues, bottled water, a
large tub of hair-removing cream. No natural light enters the room.
The window looks into the narrow well that runs through the center
of the building, but it hasn't been opened for decades and the shutters have seized together. A brand-new air-conditioning unit keeps
the room cool. One of Ariba's clients bought it because he was tired
of the heat spoiling his visits. At night Maha insists on turning it up
to maximum chill so that I wake rigid with cold under the sheet.
She's trying to be kind: she wants to make me feel at home. She
thinks it always snows in London.

Maha has new treasures she keeps safely guarded: two sets of gold jewelry, rings, necklaces, earrings, a handful of gold bangles, two mobile phones, and assorted watches. They are presents from the girls' satisfied customers, or optimistic ones who dreamed of satisfaction. It's not a large or particularly valuable horde, but Maha handles it carefully: the jewelry is displayed in red velvet boxes and the mobile phones are wrapped in tissue. She has a bank account too and has saved up almost seventy thousand rupees ($1,180). It's a start, Maha says, the beginning of a new life. "When the girls return," she says, "when Nena is married, everything will be perfect."

I've seen Adnan only once, hobbling in on crutches in the middle of the night. He was shot twice in the leg a few months ago during a gunfight in the bazaar and will probably never walk unaided again. Maha wasn't pleased to see him, judging by the way she hurled the bottle of 7-Up at his head as he left.

"I'm always alone here," she sighs. She has no customers and no husband. She spends a lot of time lying on the mattress looking agitated, fidgeting with her clothes, and smoking hashish. She's given up Corex but can't do without some kind of fix. Maha wants a man—a new one and a decent one—although Adnan would still do if he could just learn to treat her with some respect and affection and if his wife would stop concocting those vile spells. She says we both need a man and a new plan: she needs to lose some weight and I need to gain some. Today we're going on a diet.

We return from the market with two shopping bags full of diet foods: vegetables, fruit, and chicken breasts. There are two—only two—warm and tempting *naans*, wrapped in newspaper, jutting out of the bag. Maha eyes them hungrily. Bread and ghee must be eaten in moderation or not at all. A great pot of vegetable soup is simmering on the gas ring. Maha grimaces.

Charms

There's something about Maha that tells strangers she's a *kanjri*. It might be the way her *chador* comes loose and she walks around bareheaded. Or it might be the direct way she looks at men, or the way she laughs, or the way she sways through the crowd looking around with her head held high. It might simply be her beauty—lessened, but never masked, by her weight. I don't know. I'm following her through a street market on the edge of the old city. I'm about twelve feet behind her, observing the men's looks and how they whisper and nudge their friends.

The market is full of customers buying cheap clothes, household equipment, toiletries, snacks, hashish, and a variety of drugs that look like a selection of dried, crushed plants. Other traders are selling hope. Maha wants to know about the future, so we pay three rupees (five cents) to an elderly fortune-teller to have his sweet, ragged little parrot walk along a line of envelopes and pick one out with his beak. Maha isn't pleased with the fortune written on the paper in her envelope. I can't make sense of mine: it must be based on some traditional stories or part of the Quran. The old man can see my confusion, and he laughs so we can see that he hasn't a single tooth in his head.

We're going to try some more serious dealings with the spiritual world. A *jadugar*, a magician, is sitting by a wall that runs along the Ravi Road. The mystical guidance is more expensive here—twenty rupees (thirty-four cents)—but it's a more personal service. Maha has to give some details, like her name and her husband's name. She concentrates hard and throws some tiny brass dice. The *jadugar* scribbles figures on a pad and starts telling Maha all the things she wants to hear: her husband loves her but there's a problem.

"Mumtaz, his other wife," Maha interrupts, shouting so loudly that the *jadugar* flinches.

"Yes," the *jadugar* says. Mumtaz has being performing black magic on Adnan again. She's been putting her menstrual blood into his drinks. For five hundred rupees ($8) the man will produce some *tarviz*, charms that will unravel Mumtaz's spells, and everything will return to normal.

Maha settles on a fee of two hundred rupees ($3) and the man gives her three small bits of paper torn out of a notebook. There's a charm written on one and on the others there are lists of numbers in rudimentary grids. Maha folds them into tiny tight parcels and puts them in her bra. She's very pleased.

Tasneem's Lovely Hair

Malika is sitting in the telephone shop and rushes out as I stand admiring the new look of the Tarranum Cinema. It has been painted the color of dried peas. She wants to know why I haven't been to see her. The *khusra* with the harelip and the rubber legs is still in Malika's house and there's another man whom I don't recognize sitting in a corner. They're all very friendly. Tasneem is living at Malika's nowadays too, but she isn't in at the moment. They say she's gone to her village because her uncle has died. I'm not so sure this is true. Malika says that things went badly for Tasneem: her marriage ended in disaster. She says her husband was cruel. He used to beat her, and in the end, he chucked her out. She came back to Malika's because she had nowhere else to go.

She brushes my hair and asks why I'm not wearing my earrings. It's as if I'm half dressed.

A Punjabi film is on the television. Dancers run around in a field waving scarves and forming circles around a singer. The *khusras* tell me that the star is the most famous woman singer and dancer in Pakistan. The film appears very old and the heroine looks bloated and surprisingly ugly.

A *khusra* with a handsome face and a tall, slender body arrives. They laugh about how one of the other *khusras* performs her dance routines and give me a mean demonstration of her level of incompetence. The handsome *khusra* asks me how long I have known Tasneem.

"Tasneem's not good," she declares. "She's a loose person. She's *badtamiz* and useless in the head."

"She has a good heart," I add and ask why she thinks such bad things about her.

"Because she left us. We don't get married. We stay here with the others. Tasneem got married and went away."

Malika is nodding and looking at me with great seriousness. "It's not allowed," she adds. "We can't leave and get married. We can't get permission. Tasneem was bad."

Tasneem has disobeyed the rules of *khusra* society, the rules that tie them into a community and forbid them to leave, and she will be punished: she will be ridiculed and made to do all the jobs no one else will consider doing. She was treated badly before and now life will be even worse for her. I don't think it was a *badmash*, a brutal stranger, who hacked off Tasneem's lovely hair: it was the *khusras* themselves.

Jogging in Racecourse Park

Maha and I are starting an exercise regime to complement the diet. We're going to go to Racecourse Park every day. A long way from Heera Mandi, it's a park with ornamental gardens, children's playgrounds, cricket pitches, fountains, a long jogging circuit, and, in the center, a polo club. We've made an effort to look our best because it's a place frequented by the rich. The circular walk around the park is used by lots of fitness enthusiasts, and Maha joins them, walking against the flow of joggers and stopping often to admire the excellent displays of flowers and tropical plants.

Maha points out a hospital next to the park. "That's where my sister died,' she says flatly.

No one is sure why her rosy-cheeked sister collapsed last month and then died a few hours later. Not even the doctors. Maha's mother sits alone now at her window in the courtyard chewing *paan* and looking miserable. I think Maha is grieving too.

Our pace slackens to a relaxed amble. Maha is complaining: "Louise, my legs, my *bund*, my feet." She's swaying through the crowds, taking shortcuts across the grass, laughing and breathing deeply, confessing now and again that she's spotted a former client, or musing that some of the men are rich and a few of them are extremely attractive. Her *dupatta* is slung across her shoulders, a mist of perspiration dots her upper lip, and her hair, escaped from its clip, hangs provocatively around her face. Every man glances at her. A second later, they glance again.

Lahore Polo Club is in the center of the park. Maha points out all the expensive cars and says it's a good place to pick up business: she's been here before. We watch a group of children having riding lessons. Their fathers are not around: there are only a few frosty, upper-class ladies speaking English who give us disparaging looks. It's a disappointment, so we survey the car park again and lie down on the grass to watch some youths play cricket. Maha is so happy gazing at the sky and talking about how she's going to stick to the diet and be slim and meet nice men. Tomorrow she'll buy some running shoes to speed up the walking and the fat loss.

It's almost time to go home and savor some more soup, so we drag a whining Mutazar and Sofiya from the children's playground. To soothe them we stop at the cafe for everyone to enjoy ice creams and Cokes and packs of *namkeen*. Maha relishes her ice cream—after all she deserves it. A handsome young man observes every mouthful and every lick of her lips with awed fascination. He moves closer and then follows us, trancelike, to the rickshaw. Maha half-shuffles and half skips in her sandals, laughing and pulling the children along behind her. For once she's unaware of her impact, but every head has turned.

By evening Maha isn't so happy. We're cooking and she is low because I'm leaving soon.

"You are going," she complains. "There are only four days left and then I'll be alone."

"How can you be alone when you have the children?" I laugh, trying to make light of it.

"When the children go to sleep, I'm alone. My heart is alone. Those men are frauds. They use me and go away. I'll never meet a good man. Good men don't want to know me. You are the only person who hasn't given me up. Promise you won't give me up—not until we die."

I make a promise, and mean it.

A big *ghee* tin is sitting on the gas ring in the middle of the kitchen as we cook. Something is bubbling slowly inside it. It looks like volcanic mud.

"It's good magic," Maha explains. Her mood lifts and she laughs and pokes the tin. "It's got to cook every day for twenty-one days."

At the bottom of the tin, under the mud, is a little earthenware bowl. The *jadugar*'s charms are in this bowl interleaved with layers of sugar. On top of this is one of Adnan's used syringes—just to make sure the magic fastens itself on the right man.

"Adnan will smell it and know someone is thinking about him," Maha states.

"And after twenty-one days, what will happen?" I ask.

"He'll come back and love me."

Pakeezah — Pure Heart

Our rickshaw has pulled up by the barbeque stall outside the Tarranum Cinema and we're looking out of the door, shouting instructions and deciding whether to have chicken pieces or kebabs. Fewer men are around and the place has lost its bustle and energy. Many of the *kothas* are shut. I don't know whether this is because the damp, aching cold is keeping the clients away or because there's not enough custom even when the weather is good. Only a handful of performance rooms are open: they look rundown and very different from some of the lively, polished *kothas* I saw on my very first visit to Shahi Mohalla seven years ago. The shoe shops are creeping ever closer to the heart of the bazaar.

The last vestiges of the old Heera Mandi are being eclipsed and the world of the traditional brothels is almost at an end. No more *nawabs* and very few cultured *tawaifs* can be found in Shahi Mohalla. Today, an ambitious girl who wants to make it in Heera Mandi will aspire to be an actress—but most will settle for a dancing tour in the Gulf. This is where the clients are and where the money is to be made. Maha says it's good work and that her daughters can make a handsome profit. Traveling on airplanes, staying in hotels, and joining an international scene is a glamorous escape from the stigma of being a *kanjri* in Heera Mandi.

Spinning the Wheel

Maha's house is no longer half empty but full of people and noise. The girls have returned from the Gulf, and, as Maha predicted, they have brought with them a new kind of life. All are in the business. It's the kind of life Maha can remember from her own youth. The generational cycle of Heera Mandi has at last come full circle.

The rhythm of the days and nights is confusing. No one goes to bed until four or five in the morning, and they sleep until long after midday. Last night Maha said she was going to have an early night, but then she started sorting out the clothes at two o'clock. Now it's almost one in the afternoon and everyone is asleep, covered in mounds of blankets and lying perilously close to the gas fire. Four-inch flames are shooting from the top of the fire and gas is leaking and hissing around the sticky tape used to patch up the holes in the rubber supply pipe. I'm afraid there'll be an inferno, with no way out.

I leave them to sleep and I sit in the weak winter sunshine under the big shiny *panja* watching customers eating lunch on Iqbal's rooftop restaurant: they are modern, Westernized Pakistani women with short, blow-dried hair and designer sunglasses. The rough floor tiles feel warm under my feet. It's peaceful—even the horns and screeching engines of the rickshaws on Fort Road are distant sounds. Birds circle around the minarets of the Badshahi Masjid as the men make their way to prayer. In the corner of the court-yard, the topmost floor of one of the houses has caved in and a group of *khusras* edge around the chasm to sit in the sun and oil each other's hair. A couple of dogs bark at them but soon lose interest. They're always there, confined to a few feet of terrace, barking and copulating endlessly. Below me, in her own partly screened terrace, a very pretty young woman is washing dishes and throwing grains of rice to the sparrows. She has long, glossy, red nails the same shade as the flowers she grows in a pot. They are startling splashes of exquisite color in the midst of so much dusty gray and buff.

By early evening the whole family is refreshed and hungry. At midnight they are just getting into their stride. At three in the morning, it's bone-chillingly cold and the air is thick and foggy even inside the house. Smoldering hashish and fumes from the gas fires mix with the barbeque smoke drifting through the windows from the restaurant next door. They're all eating ice cream and dancing. Mutazar is throwing shoes at his sisters, shouting over the sound of the music, and ordering them to address him in the polite form of the language—as one would address one's betters. I don't feel like participating in the party and lie on the bed in a very old black coat, trying to keep warm. They think I'm ill. Nena has brought me a plate of fruit. Sofiya has offered me different flavors of ice cream, a bottle of Coke, tea, chicken curry, and sweets. She treats me like a giant doll. Whenever I'm about to sink into sleep I can hear her rushing toward me saying, "Louise Auntie needs treatment." I have pieces of wet tissue stuck all over my face.

The Old Arab

Nena is married at last. This time there was no hysteria, perhaps because she is older, no longer a child, and perhaps because the deed happened in her home, not in a foreign palace. They call her husband the Old Arab. He rings every day from Dubai to speak to

Nena and make sure she's being a faithful wife. He hasn't seen her for weeks, not since the marriage. He flew to Pakistan and stayed with them for a month. They have lots of photographs of the event: Nena in her wedding dress; Nena on the morning after her wedding; the Old Arab smoking on the rooftop; the Old Arab eating meat. They say he's 75—but he looks more like a well-used 65 to me.

Maha doesn't like him. "He took medicine," she says "so that he could have sex with Nena for hours. She was in such pain. And he has a big dick. The sister-fucker."

Nena kept pretending to have headaches and was so glad when the month was over and he returned to his car service business in Dubai.

They endured him because the money was good: five *lakh* ($ 8,400); two sets of gold jewelry, and a top-of-the-line mobile phone. He was supposed to pay another *lakh* for every month that Nena remains his wife. Maha doesn't think that this arrangement will last long. The Old Arab's already talking about halving his financial support, so Nena is starting to search for a second rich husband.

A Baby for Nisha

Nisha is convinced she's pregnant and she's delighted. The father is Azim, a thin youth with a sparse moustache and long hair that hangs in front of his eyes. Nisha has a few photos of him. She met him in Abu Dhabi when he visited her club. He's Pakistani—from Balochistan—and works as a jockey in the Gulf. Nisha has forgotten that she hates men: she loves Azim. She's deliriously happy and lies next to me on the big bed musing over whether the baby will look like Azim. Nena snorts and skips off to laugh in the other room.

Maha is sorting out the clothes again and shaking her head. She doesn't think Nisha is well enough to carry and deliver a baby. "She didn't take her pills," she complains. "And now we have this trouble." All the girls take the contraceptive pill—except Nisha who thought it would be better to be pregnant.

Maha has more to say when we are alone. "This baby is bad for us. Nisha isn't good at 'relations,'" Maha insists, "but she can dance. If she has a baby her belly will be rotten and she won't be able to dance anymore. We already have two children in the house. We can't afford another one now. It's not time for her to have a baby."

Maha understands her daughter very well: Nisha wants a baby so she won't have to work. But Nisha also wants to be in love. Azim visited her in Lahore for two days and two nights. They were the best of her life. Azim wasn't so sure about the romance. He told her, "I like you. I like your eyes and I like your smile, but I don't love you. I don't like your figure." He's already married, but Nisha is sure that the baby will be a boy and that Azim will make her his second wife.

Nisha and I share the big carved bed paid for by Nena's admirer, so I know that she cries herself to sleep every night, weeping silently for hours.

Sometimes she talks about a girl she knows in Abu Dhabi. This girl has a boyfriend, but the boyfriend doesn't like the girl's family. What, she asks, should this friend do? Should she give up her family to be with the man?

I tell her to advise her friend that men come and go but that she will always have her family. It must be the wrong thing to say because it makes her cry even more.

Last night Nisha talked with her mother and me about her future. Could she leave the family and marry Azim if he gave Maha forty *lakh* ($67,000)? Nisha is dreaming. She knows it's impossible. He'll pay nothing for her.

She's keeping a record of the relationship in a large notebook. It contains every text message Azim has ever sent her and page after page of writing in English—lines that repeat the same words over and over again: Azim I love you; Azim I love you; Azim I miss you.

Dried Fruit and Nuts

Nena has a new admirer. He saw her dancing at a function a couple of weeks ago, and he wants to get to know her better and meet her in private. We've prepared for his visit: the house has been cleaned, all the rubbish has been thrown out the window, and we have taken great care with our makeup. Sameer, exporter of dried fruit and nuts, arrives at night bearing gifts; two shopping bags full of pine nuts and sultanas, plus two truly horrendous woolen cardigans for Nena. Maha raises an eyebrow. He's pitching it all wrong. He'll have to improve the quality of his presents if he wants a piece of Nena. Sets of gold jewelry are acceptable; woolen cardigans are not.

He's an unlikely looking client: thin, 55, with flat hair and a small moustache. He sits nervously on an armchair in the main room while the family practices its craft. Nena sits next to him for a few minutes and then moves to the opposite side of the room, where she giggles, whispering to me behind her hand. Her mother and sisters take it in turn to talk to Sameer, chatting about everyday things, while Sameer casts quick but inquiring glances at Nena.

We've been eating nuts for two hours while smiling at Sameer. The nuts are fiddly to open and we're surrounded by a sea of broken shells.

Maha lies on the bed with Sameer. They're arguing.

"It's not enough," Maha is saying. "She's only been married for a month."

There's more discussion and then Maha speaks to Nena. A deal has been done: Sameer can kiss Nena and hug her—but no more. "He can't get things for free," Maha says, leaving the room with ten thousand ($169) of Sameer's rupees.

Nena is on the bed with Sameer under a big blanket. She's laughing . . . and it's time for me to go. In the other room the rest of the family is watching television. Maha joins them and settles on the mattress. She lights a joint and tears open a bag of sultanas.

New Year's Eve

Maha is highly agitated. It's New Year's Eve and the girls are performing at an important function. Everything has to be perfect. The clients are rich and influential and some of them might want more than a dancing show. The girls have been preparing for hours, choosing the right clothes, washing their hair, and applying makeup. They all want to wear bright-colored contact lenses with frighteningly blank pupils. They can't decide on blue or green and they think I'm mad when I insist they should stick with their own brown eyes—they're so much softer and prettier. Maha isn't bothered about their eyes: she's rushing around shouting about the time even though the taxi won't come for another two hours.

She looks at my makeup and pulls a face. "Make it strong," she tells me and hands me a bright lipstick. We're in matching cream silk outfits and I've been loaded up with a lot of the Old Arab's gold.

Such is the importance of the event that Adnan has been persuaded to look after Sofiya and Mutazar. I'm amazed that he's still around but he seems to be a more frequent visitor these days now that he knows Maha won't keep asking for money. He comes to take his drugs in peace, and this evening arrives with a companion, half-stoned, and with a supply that's generous enough to see them through the night. I hope the children will be safe.

The function is in a house in Defence. It's a big rambling place with an enormous hall and many half-empty rooms. A servant shows us into a room that has a bathroom en suite and the girls change into their dancing outfits. A friend of the host arrives to check us over. He has dyed black hair, sparse eyebrows, and very bad taste in burgundy, gold-buttoned blazers. Ariba says he was one of her clients a few weeks ago.

Seven middle-aged and elderly men are sitting on a shiny red-leather corner sofa in the main reception room. We sit opposite them in silence while they look at us, chatting among themselves and drinking Black Label whiskey. We're not offered drinks until a charming general gives Maha and me gin and tonics. Maha recoils at the taste.

Other guests arrive and we're shunted down the room to sit on less expensive chairs farther away from the food and drinks. One of the guests is doddery, fat, and covered in the blotchy pigmentation of the extremely old. He eases himself into the corner of the sofa with difficulty.

"Give me drink," he calls to Nena. She walks across the room with perfect poise and proceeds to pour an entire tumblerful of whiskey. There are roars of laughter—she looks so naive. For a second, Nena loses her composure. As she stares at her mother in panic, Maha fixes an intent gaze on her daughter, smiles, and tells her that it's okay.

"She's very young," she explains to the *tamash been*. The old man pats the sofa and motions Nena to sit down. He plays with her fingers and places a hand on her thigh and beneath her carefully painted smile—the one she spends so long cultivating in front of the mirror—Nena is rigid with humiliation. Men never touch women this way in public in Pakistan.

The room is filling up with rich men and their "keeps," their "kept women." This isn't a party for husbands and wives but for the male elite of Lahore and their current mistresses. The men are industrialists, bureaucrats, generals, and senior professionals. Their keeps, beautiful women in their twenties and early thirties, are anxious to differentiate themselves from the dancing girls of Heera Mandi. This is probably because most of them also have their roots in the *mohalla:* they or their mothers were *nachne walli* too and were lifted out of the brothel to become the mistresses of influential men. They treat us with complete disdain and pretended indifference. When the music begins and Nena starts to dance in the center of the room, all the men are captivated but their women feign boredom. The host's mistress, a well-groomed but hard-faced and thin-lipped woman, casts her eyes to the ceiling.

"Bitch," Maha mutters. I couldn't agree more.

Nena's performance is accomplished and she's far more beautiful than anyone else in the room. All the women are aware of it. She hides her face with her hands and peeps between her fingers. She stamps her feet, falling to her knees, arching her back so that her silky black hair cascades onto the floor. The *tamash been* cover her with money as she spins. They hold thousand-rupee bills next to their friends and Nena dances over to take the money and drop it on the carpet. Maha is keeping a keen eye on the servants to make sure they don't pocket the family's profits.

The audience is calling *"Vah, Vah!"* (Bravo) in appreciation and the bitter keep is furious. When the music ends she glides passed Nena and presses several buttons on the music system. It's such a sophisticated but badly put-together system that it takes a long time to reset. Nena has been sabotaged.

Once the music has been restored, the other girls take turns dancing: Nisha does so stiffly and Ariba performs in a white glitter *shalwaar kameez* that makes her large breasts look even bigger. One of the generals is riveted.

He wants to meet her but doesn't get the chance because we are asked to leave immediately after midnight. The thin-lipped keep insists on it. Perhaps she doesn't want to be categorized with the fallen women of Heera Mandi or perhaps she doesn't feel comfortable about the way her own patron is looking at Nena.

"Insults. Insults," Maha fumes as we stuff all the dancing outfits into the suitcase.

The blotchy old man shouts at us to go quickly and laughs as the servants usher us through the doors.

"Our honor is destroyed," Maha moans. "That bitch woman. I hope she dies."

We're in the taxi, driving home in silence. Nena's forehead is pressed against the window, Nisha is crying, and Ariba has her head in her hands. Everything has gone wrong. We have the money, barely contained, inside Maha's handbag, but the family's honor has been compromised. All those *tamash been* will know that we were humiliated. It'll do the girls' reputations serious damage. Just as important, Maha and the girls feel that their pride, that thing they cling to and preserve through good times and bad, has been injured.

Maha wipes her eyes on her *dupatta,* rocks back and forth in her seat, and tells the taxi driver to go to Mall Road. We stop outside a drive-in restaurant and Maha orders plates of fried chicken, french fries, and so many bottles of Coke that we're soon stuffed with food and everyone is beginning to laugh at the memory of the keep and her acid face.

To add to the fun we ask the taxi driver to stop at a bakery on the way home. We watch Maha though the windows, pointing to this and that.

"Look at Mum," Nisha says, "she's buying the whole shop."

Maha pulls thousand-rupee notes from her bag and the shop assistants carry *shoppers* full of food to the car. At home, the food is spread on a *chador* in the best room. We have an odd combination of olives, *namkeen,* dates, biscuits, tinned pineapple, and a giant ice-cream cake. Adnan has woken up and looks on sleepily as Maha sticks a dozen candles into the cake so that we can celebrate New Year in style. We all sing "Happy Birthday to You" and Maha laughs, watching the children blow out the candles. But then she roots around in her bag and shouts at the girls to bring the Old Arab's phone. They can't find it and no one can remember seeing it since we left the function.

Maha has stopped laughing. She's stricken. The phone is worth a fortune and all the clients' names and phone numbers are stored on its memory. It is, as Maha often reminds me, her "big pimp." Losing the database will be a disaster. "What a night. What a night," Maha is crying. The ice cream is melting fast in front of the gas fire, but no one cares.

We think it might be in the taxi, so we rush out into the street and run through the bazaar. The taxi *walla* is pulled out of a tea shop. The phone isn't in his car, but Maha has made up her mind that he stole it while she was buying the food in the bakery. Adnan talks to the driver and asks him to come to the house, where he sits looking patient as Maha shrieks that he's a thief and that she'll call the police. Ariba adds her own loud voice and Nena and Nisha look on accusingly. "You stole my phone because your wife is pregnant and you need the money," Maha maintains. The taxi driver just shakes his head and leaves.

The next morning the phone still hasn't been found. Maha is catatonic and Nisha is fretting that she'll be missing text messages from Azim. One of Maha's many cousins is here to discuss the crisis. I can hear them talking in the other room and deciding what to do.

"Shall we ask Louise what she thinks?" the cousin suggests.

"No," Maha answers distractedly. "She's simple. She won't understand."

An hour later Maha has her phone back. The taxi driver said it had fallen down the side of the passenger seat. It's not true. We checked there last night and when we dialed the phone's number from Adnan's mobile it had been switched off. The driver must have stolen it but then relented under pressure.

Maha has collapsed on the bed with relief. Nisha is sad: Azim hadn't sent a message after all and Nena is wondering if the Old Arab will be angry because she wasn't on the other end of the line when he called. The reflections don't last for long because there's a loud shout from the family living on the ground floor. Maha has left the water pump on and a flood is cascading down the dark well of the building. Nena shouts back and a volley of abuse follows from the depths of the house. Nena replies with uncharacteristic venom and Maha joins in with her usual gusto.

"You've a fire in your cunt," the downstairs woman calls.

Maha roars as we drag her back from the railings. "That bitch has two hundred men a week."

Dreams of Abu Dhabi

Ariba has put on a lot of weight. She looks 30 not 15. The clients must like it because plenty of men request her services. She's no

longer the louse-ridden girl of two years ago—but she's still "bad."
She smokes in the bathroom and makes no effort to hide the evidence. There are cigarette butts floating in the toilet and a box of
matches on the long-defunct cistern. She drinks alcohol with the
clients too. Elite dancing girls are not supposed to drink—it's
shameless—but Ariba isn't an elite girl. She's a fun girl and, though
she earns far less than the beautiful Nena, she has a lot more customers. She keeps their phone numbers in a tin: some have given her
their business cards and others have left their names scrawled on
scraps of paper, tissues, or napkins.

Her success hasn't erased her browbeaten look and she's just as
desperate for affection. She's given me dozens of little gifts: hair
clips, hairbands, eye shadow, socks, bubble gum, bracelets, a picture of herself. When I apply some makeup she stands close by
watching intently and breathing heavily. She sits next to me and
stares when I try to work. She'll sit quietly for hours, just looking. I
can't believe I'm so interesting.

Her sisters make an effort to look glamorous around the house,
but Ariba never bothers. She slops around in red tracksuit pants and
an old lamé top. She doesn't even change when *dalals* come to the
house. Today a big *dalal* who supplies girls to senior bureaucrats and
industrialists has called to assess the girls informally. He's in cricket
whites, a blazer, and a knitted woolen hat and relaxes on the bed eating pineapple cake. The girls are courting him. He's very funny and
makes Maha laugh. They're discussing Nena and the *dalal* advises
her not to wear too much makeup: she is young and fresh and has a
natural beauty. Many *tamash been* value such beauty, he adds with a
nod to her mother.

Ariba hasn't joined us on the mattress. She's busy cleaning in another room. She never stops working; she's always peeling vegetables, washing dishes, or sweeping the floor. Whenever she sees me
with the *jharu* she snatches it from my hand and shouts, "No, Louise
Auntie. Sit down." I'm tired of sitting down, watching her work.

She works hard at pleasing her clients too. Tonight she's entertaining a customer at a big hotel. She'll spend two hours getting ready. He lives in England and always asks for Ariba when he comes to Pakistan on business. He's become a regular and Ariba says he's okay. She knows what he wants: a night full of cuddles and kisses because he can't manage much more.

"What do you wish for in your life?" I ask her as she's getting ready.

She looks at me as if it's a really odd question. "I don't know."

"You want nothing?"

"A car, perhaps. Some nice clothes."

I look at her quizzically. She's nervously excited. "Perhaps I will go back to Abu Dhabi." She tries to look relaxed and unfazed by the topic, but it's not convincing.

"You like Abu Dhabi?"

"Yes."

She stops applying the makeup and shows me a photograph she keeps hidden in a zipped compartment of her bag. It's of a very ordinary middle-aged man. He's standing, slightly awkwardly, in a dry-cleaning store, in front of racks of laundered dresses and suits, and he's smiling. It's a gentle, embarrassed, bashful smile. Looking at him makes Ariba smile too.

I ask her why she likes Abu Dhabi and she struggles to find an answer. There's a long pause and then she sighs deeply. "Because here I am old and there I am young."

Behind Locked Doors

Everyone is irritable and complaining of stomach cramps. In households of women, reproductive cycles synchronize and everyone has her period at the same time. Customers aren't welcome for a while. Nothing is happening. Life is on hold and there's no business to structure the day. We sleep most of the time. There's no difference between day and night. Sameer, exporter of dried fruits, phoned but Maha was frank with him. When he spoke to Nena, he said he'd call back in a week once the "menses" were over.

We're waiting to hear confirmation of Nisha's pregnancy, anxious for news, but she's said nothing. She's quiet and withdrawn over the meal, and then Nena points to some spots of dried blood on Nisha's *shalwaar.*

"It's from my nose. It's from my nose," Nisha screams and rushes from the room, not knowing whether to clutch her face or her *shalwaar.*

She's sobbing quietly on the bed. There is no baby and there will be no marriage.

We were woken by an enormous crash at dawn; we'd only just gone to sleep. The storage unit in the kitchen had fallen over. Shattered glass covered the floor and Maha's spices, sugar, and salt lay in untidy piles. The tomato ketchup looked like blood. We were all too tired to bother to clean it up, so we left it: we'll deal with it like the rest of the mess—when the customers return. We'll also have to deal with the aftereffects of a flood from the bathroom. Mutazar or Sofiya must have left the tap on and the water spilled into the hall, on to the carpet, and down the stairs. Ariba mopped it up with one of Sameer's dreadful cardigans. It's been lying, sodden, in the hall for three or four days. It's icy when you step on it and the wool has matted into a dense green mass.

Maha's mother is weeping in the other room. She's heard that Maha doesn't want her to visit. It's true. Maha is scared. Things aren't going well again. There was the humiliation of the New Year's function, the phone crisis, the toppled cupboard—and Maha is convinced it's because of her mother's black magic. She's jealous of their newfound prosperity. I'm not surprised; Maha doesn't share her daughters' riches with her mother. Blood was sprinkled on the stairs last week and we all trampled it into the house. Maha says her stepfather is responsible for the worst of it.

He took soil from the graveyard and added it to the blood on the steps. We're living amid powdered death.

S ameer is on the phone again today and Maha is talking softly and laughing, calling him *beta* — son — while chiding him that he's not offering enough. She's speaking about insults and honor.

"Look," she says in exasperation, "this is a business and I'm not giving my daughter for free."

She turns to me and shrieks, "He's bringing us some more nuts and dried fruit."

Nena takes the phone and giggles. Her mother is livid but faintly amused at the man's stupidity. "He thinks he can pay for my daughter with five hundred rupees' worth of dried fruit and nuts. Idiot."

A riba's been thieving again. Maha found lots of my things in her cupboard — nothing important or expensive, just little things like mints, felt-tip pens, a pink lip gloss. She's also been using my mobile phone to call someone in Abu Dhabi. Maha is furious. And she thinks Ariba is stealing more than mints: she thinks she's keeping money that the clients give her instead of handing it over to the family. Ariba sobs and cowers in the corner, her glass bangles broken and blood trickling down her arms. Maha is screaming at her. "Your father was a dog and you are a dog."

Maha's breathing is jagged and shallow, her eyes filling up with tears as she sinks onto the bed. Her anger at Ariba's financial independence is rooted in fear. "My daughters don't want to live with me," she cries. "They want to leave me." Maha is afraid they will do what she did to her own mother and that she'll have no one to support her as she enters middle age.

Ariba is saved from another beating by a phone call from the Old Arab. Nena doesn't want to talk to him, so Nisha takes the call. "Hello, my darling," she breathes. "I miss you. When will I see you?"

The Old Arab can't distinguish between his wife and her sister.

Nisha covers the mouthpiece to tell us that the Old Arab is drunk. We all have a turn listening to him slurring words of love and lust. "It's a dirty call," Nisha explains, without bothering to hide her hysterical laughter.

Nena rolls her eyes and watches television while Nisha remains half engaged with the call, eating *namkeen* and remembering, now and again, to ask the Old Arab what he'd like to do and to murmur, "My darling." It's a hopelessly amateur performance, but the Old Arab is too drunk to care.

Pictures from a Brothel

Maha's diet is working. She's far slimmer and can wear many of the clothes she hasn't been able to squeeze into for years. She's happier: happy that they no longer have to beg for money to pay the rent, happy that her daughters are successes. Adnan makes a little contribution to the house but it doesn't really matter now that they have other sources of income. They will be secure while the girls' youth and beauty last. Adnan must be happy too: he can enjoy his drugs in nicer surroundings, paid for by his stepdaughters' labor.

Maha is telling me that she wants to come to England. She wants to come to my wedding when I find a husband—as she's so sure I will. Then, after the wedding, she wants me to bring him here to Heera Mandi.

"Promise me, Louise. Swear on the Quran. Promise Shahbaz Qalandar: you will bring him here."

Maha insists that I'll sleep with him on the big bed and that she's going to treat him like a king and cook him stuffed *karela*—a short, knobbly-skinned squash that she packs with lamb and spices and fries in a delicious gravy. It will make him strong. They used to give it to all the prestigious clients, but today they don't bother. The clients just bring their own Viagra. She says no husband of mine—no man married to her own sister—will ever require medicine. It's the greatest compliment she's ever paid me.

I can never thank Maha enough for all she's given me: a window on her world and her generous, all-embracing, sometimes overpowering friendship. I go home to England soon but these won't be my last days in the *mohalla*. I'll return in the summer and, like Maha and the rest of the *nachne walli*, I will never really leave Heera Mandi.

I treasure Maha for all the things she is and try to understand all the things she isn't. She has moments of such spectacular anger and cruelty, and at times these are almost unforgivable. I pity and admire Ariba, the loveless daughter, but I am sad too for Maha—born and raised in a brothel and sold as a child. Such violence breeds violence. Maha should have had all the joy of life knocked out of her. She should have no affection left for her children. She should be the most strident, embittered cynic. And she should hate men—but she doesn't: she still, miraculously, believes in love. Maybe that great romantic hope is the thing that keeps her going, even though all around her she can see that love is a chimera, that it can never last, that the women here are playthings, valued only for their beauty and erotic skills. One day, she's convinced with all her heart that she will meet a man who will love more than the dancing girl.

After a long break Iqbal is painting again. He's working on Maha's portrait: pictures of her dancing; of her sitting in a shaft of afternoon sun; of her with the children. She's flattered by all the attention.

"Make me look small," she instructs. "Make me look slim."

He laughs and tells her to be still for a moment so he can begin to capture a fragment of her beauty and her restless energy. He adjusts his easel and mixes his oils, but she's already moved again. He smiles patiently and sighs, exasperated and yet captivated by his model. He won't be able to make her look small. It's impossible. He couldn't make her look anything other than larger than life.

The Triumphs of a Nachne Walli

Nena is very different from the shy and gentle 12-year-old girl I met almost five years ago. She's no longer a child but a young woman. She also has a harder edge and an unwieldy ego. She tells me over and over again about her triumphs with men and her marriage to the Old Arab. She's full of herself: confident of her beauty and seductive powers. She spends several hours a day applying makeup and preening in front of the mirror. At the moment she's particularly taken with applying a special kind of lengthening and thickening mascara that makes her lashes look as if they're coated in iron filings.

She's sitting with Nisha and me on the bed. Maha has gone out to see the magician. I'm hearing a familiar tale about how all the men in the club in Dubai are enchanted by her beauty. She mentions one name again and again and it's not the Old Arab's. She closes the door.

"Louise Auntie, I don't want to have relations with old men," she whispers seriously. "I want to have relations with young men."

"Only one young man," Nisha adds.

Nena grows flushed and embarrassed.

"Who is this young man?" I ask.

Nena can't speak, so Nisha answers. "Yusuf. He's an Arab. A young Arab from Dubai."

Nena is nodding and now she can't stop speaking. "He's eighteen and handsome. So very handsome . . . and I love him."

"He loves her too," Nisha adds conspiratorially. "And his family is rich."

Nena has been out with him in his car. They went to the beach and he kissed her, and it was nothing like kissing the Old Arab. It was beautiful. She shivers at the memory.

"He wanted to marry her but his family wouldn't allow it," Nisha chokes, thinking of her own bad luck in love.

So, instead, Nena married the Old Arab. No wonder she has a

harder edge. Now she can never be Yusuf's wife, or the wife of any respectable man. Maha thought it was better this way—she didn't want Nena falling in love with a young man, surrendering her virginity for free and then leaving the family. The boy would only abandon her in the end. Nena has promised her mother that she'll never see Yusuf again. I don't think she will keep this promise: one day, even if only for a day, she'll be with Yusuf. And while she waits to enter this heaven, she'll apply her makeup and smile and dance for other men as if it were her joy and her whole life.

Pakeezah

We are sitting in the best room talking about my work and the book I'm writing about Heera Mandi. Maha wants to know what new things I'm saying about her. I say, "Everything," and she's pleased.

"I'm the star of the book, aren't I?" she questions.

I confirm that she is and that the children are stars too.

I think I understand the kind of star Maha wants to be. She enjoys lots of Bollywood films and she knows one especially well— *Pakeezah*, which means "Pure Heart," a classic film made in the early 1970s. It's a story about a *tawaif* who is rejected by her lover's family and who dies in childbirth. Her daughter, a courtesan too, struggles for honor and fulfillment. The film romanticizes the world of the *tawaifs* even as it damns it. Meena Kumari, the legendary Bollywood actress, played both the lead characters, alluring and gracious even as she neared death both in the film and in reality. Sumptuously dressed, adored by men, technically skilled in the performing arts, innocent and yet battered by life, the courtesans in *Pakeezah* possessed the pure heart of the film's title. It was Meena Kumari's last movie—arguably her most famous—and one that immortalized the tragedy of the courtesans she played and her own tragic death from alcohol-induced cirrhosis of the liver. Perhaps Maha wants this kind of immortality too—a lasting record of her life, something to lift her out of the ghetto.

I explain that I've got to change all the names in the book so no one will know for sure who they really are.

They sit around and argue about how they would like to be known by people in America and London. They suggest first one thing, then another and then it's settled. I write their pseudonyms on a piece of paper and they look at it, seeing how the names appear when they are written in English. They're unusual names in Shahi Mohalla: they've chosen the names of higher-status women—the women they'd love to be.

I've been in town all afternoon sorting out my return ticket. It's uncharacteristically quiet as I walk up the flights of steps to Maha's house. The children aren't fighting, I can't hear Maha's voice, and the doors are all open.

Sofiya and Mutazar are bending over a large cake. Bits of pineapple have fallen on to the mattress and cream is smeared all over their faces. A chocolate ice cream cake is half defrosted and oozing over the rug.

"What's happened?" I ask. "Who bought the cakes?"

"A *tamash been*," Sofiya mumbles through a mouthful of pineapple.

"Sameer?"

Nisha and Nena shake their heads. They aren't happy. They're standing stiffly at the far end the room. They don't like what's going on: they thought that they had eclipsed their mother at last.

Maha is sitting on the bed I share with Nisha and she beckons me to come quickly. There must be four dozen roses in the room. They are tender-stemmed, pink, red, and yellow, picked just as the buds are about to open into full bloom. Mutazar and Sofiya have stuck each one to the walls with bits of sticky tape. It's strangely, surreally beautiful.

The flowers and the cakes were delivered to Maha from an admirer who chose not to give his name. A card came with them. The printed verse is written in English and I struggle to translate it slowly and hesitantly into Urdu. It's about the path of life, and tears and a lost heart and a soul that finds the place in which it belongs — a place in which it will find happiness. Maha listens intently, staring at the card. And when I've finished my clumsy translation she makes me repeat it.

She turns to the wall and places her hand over one of the flowers. It's already beginning to wilt.

Maha smiles and presses her head against her hand. "Tell me again," she sighs. "And after that, tell me again."

On my last day in Heera Mandi I've been asked to bring a notepad and pen into a tiny interior room full of suitcases and old clothes. Maha is pulling her hoard out of a steel box. She's taking it to the bank, but she wants to count it all first. Rolls of notes, dozens of mobile phones, watches, fancy electronic gadgets, and boxes of jewelry are stacked in neat order. We sort the money into piles: the five *lakh* ($8,400) from the Old Arab — the price of Nena's virginity — is arranged in neat rows all over the floor. The money from the dancing tours is sorted and arranged in another careful row. Nena has totted up the total, and I have to check it to make sure they've got their sums right.

It's a small fortune but one that can't be earned quickly again — not until Sofiya enters the business. Maha will have to be careful how she spends the cash. It's the family's future and its security. They're going to use the money to put a deposit on this house so that, one day soon, they can buy it and stop paying rent.

We wrap up the hoard and put it in bags that are strapped to Maha's body and concealed by a giant *chador*. We stop by the door to put on our shoes and Maha pauses. Someone is coming up the stairs. They're dragging something.

It's Adnan. We can hear him pulling his crippled leg behind him.

We look down the stairs. A man is here, but it's not Adnan. It's Sameer, exporter of dried fruit and nuts. He's puffing his way toward us heaving a sack.

"What's that?" Nisha asks as he drops the sack at Maha's feet.

We're staring at his expectant face, his moustache spiky with excitement and effort.

"Pistachio," Sameer crows triumphantly.

Nisha pushes open the door to the best room and Nena vanishes to put on her makeup.

"*Beta,*" Maha chortles as she kicks off her shoes. "Come in."

Afterword

It has been one of the greatest privileges of my life to write *The Dancing Girls of Lahore*. As an academic at the University of Birmingham in the United Kingdom, I've spent years researching prostitution and the trafficking of women and girls throughout Asia: the Thai and Filipino girls working in Japanese clubs; the sex-tourist venues of Pattaya and Angeles; the sexual exploitation of children in Cambodia's shack-lined red-light areas; and the migration and trafficking of teenagers from Nepal's beautiful mountains to the vast brothels of India's congested cities. But nowhere has been quite like Heera Mandi, with its long artistic traditions, its strongly felt Muslim religion, and its sense of community, tightly bound with Shia ritual. In the late 1990s it was clearly a community in transition, moving swiftly from an old-world brothel district steeped in artistic performance and the romance of purchased love to a more modern red-light area, stripped of elite pretensions and reliant upon the sale of sex. For an academic, it offered unbelievably rich material for research.

This research has not always been easy. When I began working properly on a study of Heera Mandi in early 2000, I had three young children whom I couldn't take with me to Pakistan. As they have grown, it has been even more important for them to remain at home in the comparative safety of an English suburb. The contrast between the lives of my two teenage girls and those of Maha's daughters has often struck me forcibly and with an inescapable guilt, because while 14-year-old girls in my middle-class family in Britain go to school and to the cinema, Maha's pretty daughters, the same age as my own, dance for men and have their virginity purchased by the highest bidder. And while Maha and her children sell sex, I live, at least in part, by teaching about what they do.

This book is taken directly from the daily diary I wrote while undertaking academic research that was partly funded first by the Nuffield Foundation and then by the British Academy. My thanks to both these institutions for their generous support.

As every author knows, books are the work of many minds and the unrecorded labor of others, and *Dancing Girls* is no exception. By right, a long list of names should appear alongside mine on the book's cover. My mother and father, Julie and Peter Brown, looked after my children, uncomplainingly, while I returned repeatedly to Heera Mandi over four and a half years. Without their constant backing I could not have begun this work, never mind completed it. In Lahore, I was taught Urdu by Naveed Rehman, who, I am sure, despaired of my bad grammar and linguistic idiocy. The swear words that appear so often in this book were not the result of his tuition, and I thank him for his long suffering. My agent, Caradoc King, and editor, Courtney Hodell, have both had a profound influence on the book, prodding at weaknesses and building on its strengths. I am grateful to them both for their vision and kind encouragement.

Finally, and most importantly, I want to thank all the women and men of Heera Mandi. I thought I came here to write an academic analysis—and eventually I will do this—but this place and its people have touched me in a way that is as much personal as it is analytical. I have learned so much about the capacity to love and be joyous in the midst of cruelty and social stigma. Occasionally I have been scared here, sometimes frightened by the *tamash been,* but never by the people who work and live in this *mohalla* and who have been, year after year, consistent in their warm friendship to an odd outsider.

Iqbal, I thank you for our thoughtful, languid, funny evenings on your rooftop. Tariq, you have a family to be proud of, and Tasneem, wherever you are, I thank you for the purple suit and the sparkle that so often broke through your layers of worry.

Maha, Nisha, Nena, and Ariba, what thanks can I write for you, when words will never be enough?

Glossary of Urdu and Punjabi Words

ajayabghar   Museum

alam   Copies of the battle standards used by the supporters of
Hussain at the battle of Karbala

andron shaher   Walled or old city of Lahore; inner city

ashik   Lover

azan   Call to prayer

badmash   Gangster, villain, thug

badtamiz   Stupid, without sense

bedagh   Name given to passive male sexual partner in parts of
northern Pakistan

beta   Son

beysharam   Shameless

bhut   Ghost

booti   Plant-derived narcotic

bund   Ass

burqa   Traditional long-sleeved gown that covers women from
head to foot; in Lahore these are usually black; often worn with
a *nikab*—a veil covering the face

chador   Large sheet or veil used to cover the head; also used as a covering for other things, such as a tablecloth spread on the floor at mealtimes

chae walla   Man who makes and sells tea

chakla   Traditional red-light areas of South Asia

Chand Raat   The night of the new moon; the beginning Eid-al-Fitr

charpoy   Traditional wood-framed, rope-strung beds; modern versions have metal frames and are strung with nylon

chela   Apprentices, disciples

chota, choti   Small, little, young

chowk   Intersection

churha   Group of untouchable people in traditional South Asia

dai   Traditional midwife; in Heera Mandi she may also perform abortions

dal   Lentil soup eaten with bread or rice

dalal   Agent, promoter, pimp

darzi   Tailor

deg   Metal pot or cauldron

devadasi   Temple prostitutes

dhobi walla   Washerman

dholak   Double-sided drum

dhoti   Piece of material that men wrap around their waists like a skirt; usually reaches to mid-calf level

dubanna walla   Masseur practicing a form of massage in which the limbs are squeezed

dudh walla   Milkman

dupatta   Shawl or scarf used to hide the hair and breasts

gali   Lane

ganda, gandi   Dirty

gandi kanjri   Insult meaning "dirty prostitute"

gashti   Insult meaning "Low-class prostitute"

geet   A form of popular song

ghazal   Songs in a classical or semiclassical tradition

ghee   Clarified butter used extensively in cooking

ghungaroo   Ankle bells worn by dancers

gola   Ice, popsicle

goree   White woman

gulab   Rose

hakeem   Traditional healer

halva   Sweet made with semolina, sugar, and flavorings

hath ka matam   Self-flagellation using the hands to beat the chest

haveli   Large traditional houses built around a central courtyard

hijra   Person possessing both male and female characteristics

hookah   Large upright pipe and tube through which tobacco is smoked

inshallah   "God willing"

izzat   Honor

jadu   Magic

jadugar   Magician

jharoka   Traditional, carved, wooden bay windows

jharu   Broom made by binding twigs together

kachumber   Salad made with tomatoes, onion, cucumber, and lime

kafir   Non-Muslims

kala ilm   Black arts; knowledge associated with demons and black magic

kameez   Long shirt or tunic; worn by both men and women

kanjar   Traditionally, a caste of entertainers

kanjri   Prostitute

kathak   Dance form

kharab   Bad, spoiled, rotten

khusra   Person possessing both male and female characteristics

kofte   Meatballs

kotha   Performance rooms associated with the sex trade

kothi khana   Cheap brothels with male pimps

kusi   Vagina, cunt

lakh   Number equivalent to 100,000

lathi   Long bamboo cane

lun   Penis, dick

majlis, majalis   Gatherings to commemorate the suffering of the Shia martyrs

malish karne walla   Masseur

mallik   Landlord

masjid   Mosque

matam   Self-flagellation

maulvi   Holy man, preacher

mazar   Tomb, mausoleum

mehbub   Lover

mehindi   Henna; designs, painted on face and hand

mela   Festival

mirasi   Musicians

mishar   Water carrier

mohalla   Neighborhood

mooli   Long white radish

motiya   Small, white, highly fragrant flowers

muezzim   Person who calls the men to prayer

mujra   Traditional dancing and singing performance

mullah   Holy man, leader, teacher

mut'a   Temporary contract marriage

naan   Bread made with yeast

nachna   Verb meaning "to dance"

nachne walli   Dancing girls

naika   Brothel manager

namkeen   "Bombay mix," a spicy fried snack made from wheat, nuts, and dal

nawab   Wealthy landowner

niche walla   Person from downstairs

nikah nama   Legal contract of marriage

nuri ilm   Luminous knowledge

paan   Traditional and popular stimulant made from tobacco, betel nut, and spices wrapped in a betel leaf

paandan   Box containing the ingredients for making *paan*

panch ab   Five Rivers; the region from which the name "Punjab" derives

panja, panje   Protective Shia hand; Shia symbol

paow   Stewed goats feet

paparh   Pappadam; a fried snack

patake Firecrackers

pik Red-stained saliva produced by chewing *paan*

pir Holy man associated with the Sufi tradition

purdah Practice of secluding women

puri Fried bread

qawwali Music of Sufism, Islam's mystical tradition

rickshaw walla Rickshaw driver; rickshaws are motorized and carry two to three passengers

roti, rotiya Thin, unleavened bread

sabil Water stall

saf Clean, pure

salan Gravy in which curries are cooked

samosa Deep-fried pastry snacks filled with spiced vegetables or meat

sanam Lover, beloved

sarangi Small string instrument common in traditional northern Indian music

sayeed Muslim who can trace his lineage to the family of the Prophet

shadi Wedding ceremony; also used as a way to describe a transaction between a sex worker and her client

shalwaar Traditional baggy trousers worn under a kameez; worn by both men and women

shalwaar kameez Traditional dress in large parts of Pakistan and Punjab; consists of trousers and a tunic or shirt

sharif Noble, respectable

sherbet Traditional drink made from sweetened, diluted fruit juices; often now made with artificial favorings

shohar Husband; in Heera Mandi it also means partner

shopper Plastic shopping bag

sita Corn cob

surma Kohl

tabla Pair of small drums; an essential instrument in South Asian music

talaq Divorce

tamash been Customers; technically spectators who watch a performance but don't take part

tanga Horse-drawn cart

tarviz Charms, amulets

tawaif Respectable name for a prostitute, similar to a courtesan

taxi Extremely derogatory name for a prostitute

tazia Models of tombs used in Shia religious rituals

thanna Police station

topi Small hat

upar walla Person from upstairs

urs Commemoration associated with the anniversary of a death

walla Person linked to an occupation or place

zanjiri matam Self-flagellation with blades strung on chains

zenana Living quarters set aside for women who are secluded

zina Unlawful sexual intercourse; in practice all sex outside marriage

ziyarat Participation in religious ritual

Index

Index

Balochi language, 134
bamboo canes (*lathis*), 54, 105
Bangladesh:
  brothels of, 8, 188
  Heera Mandi families from, 149, 181–83, 200
beds, rope-strung (*charpoys*), 7, 8, 32, 34, 73, 74, 81, 112, 114, 129, 131, 145, 146, 161, 181, 182, 188, 193–94, 197, 205
Begam, Nadira, 30–31
beggars, 16, 72, 79, 136
Best Musical Group, 198
betel leaf stimulant (*paan*), 7, 13, 42, 44, 99, 130, 141, 178, 187, 202, 266
Bhati Gate, 37, 131, 132, 163, 169, 188, 207
Bilquis, 181–82
bin-Laden, Osama, 241
black magic (*kala ilm*), 86–90, 174, 225, 264
blindness, 145
Bombay, 31
bras, 82, 121
  as "Pakistani pockets," 140–41
bread:
  fried (*puri*), 4
  leavened (*naan*), 24, 151, 263
  unleavened (*roti; rotiya*), 10, 22, 48, 87, 108, 151, 173, 178
British raj, 3, 5, 145
  British women imported as wives for, 36
  cultural changes under, 35–37
  1857 mutiny against, 36
  independence of some princely states from, 36
  Indian women taken as concubines by, 35, 36
  1947 departure of, 42
  social distance encouraged between natives and, 36
  Victorian standards of, 36
brooms (*jharu*), 130, 280
brothels:
  in Asian cities, 8, 291
  drug use in, 89
  of Heera Mandi, 1–2, 5, 8, 12–13, 23, 28, 40–42, 45, 51, 56, 62–64, 70–72, 259, 269–70
  in hotels and suburbs, 41, 126
  management of, 45, 51, 68, 70, 133, 135, 143, 166, 209
  sexual slavery in, 45, 291
Brown, Louise (author):
  academic research of, 7–8, 291–92
  atheism of, 106
  divorce of, 57–58
  Maha's relationship with, 93–95, 100–2, 106, 107, 119, 131, 169, 173–74, 180–81, 205, 212, 225, 284–85, 293
  physical appearance of, 16, 119, 128

teaching of, 8, 93, 189, 291
three children of, 8, 16, 93, 119, 162, 291–92
Burton, Richard, 96
buses, 95, 96
butchers, 4, 207–8
butter, clarified (*ghee*), 16, 93, 160, 263, 286

Calcutta, 8, 188
call to prayer (*azan*), 7, 154, 204
caste system, 160
  entertainers in, 27, 35
  prostitutes in, 27–28, 35
  subcastes in, 27
  untouchables in, 77–80
castration, 31, 46–47, 49, 50, 124
cauldrons (*deg*), 37, 159–60, 199
Chand Raat (Moon Night), 147
charity, 159–61, 164
children:
  clothes of, 108
  eating hierarchies among, 22, 24
  fatherless, 155
  sexual exploitation of, 8, 183, 291
  status of, 17, 22–24
  support of, 60
Children's Illustrated Bible Stories, 102
Christians, 88, 245
  alcohol licenses granted to, 75
  Christmas celebration of, 144–47
  Pakistani, 75, 76–80, 144–47
  prostitutes as, 37
  untouchables converted by, 78–80
Christmas:
  celebration of, 144–47
  symbols of, 127, 128, 146–47
Christmas carols, 146
cinema, 11, 46, 186–87, 292
  mobile, 39
  see also Indian films; Pakistani films; Punjabi films
circumcision, 49
circuses, 125–26
concubines:
  in hareems, 30–31
  Indian women taken by British as, 35, 36
condoms, 134–35
contact lenses, 123
contraceptive pills, 273
cooks, 160, 199
Corex (cough medicine), 136–37, 138, 141, 143, 152, 178, 263
courtesans, 9, 13
  affluence of, 59–60
  aristocrats and royalty as clients of, 5, 27, 31
  careers of daughters and nieces managed by, 18, 19

Index

Index

Index

Index